THE BEDFORD SERIES IN HISTORY AND CULTURE

Women's Magazines
1940–1960

Gender Roles and the Popular Press

Edited with an Introduction by

Nancy A. Walker

Vanderbilt University

Palgrave Macmillan

For Bedford/St. Martin's

History Editor: Katherine E. Kurzman
Developmental Editor: Kate Sheehan Roach
Production Editor: Stasia Zomkowski
Marketing Manager: Charles Cavaliere
Production Assistants: Melissa Cook, Beth Remmes
Copyeditor: Barbara G. Flanagan
Text Design: Claire Seng-Niemoeller
Indexer: Steve Csipke
Cover Design: Richard Emery Design, Inc.
Cover Art: Women under hairdryers reading. UPI photo, 1961. UPI/Corbis-Bettmann.
Composition: ComCom
Printing and Binding: Haddon Craftsmen, Inc.

President: Charles H. Christensen
Editorial Director: Joan E. Feinberg
Director of Editing, Design, and Production: Marcia Cohen
Managing Editor: Elizabeth M. Schaaf

Library of Congress Catalog Card Number: 97–74972

Copyright © 1998 by Bedford/St. Martin's

Manufactured in the United States of America.

2   1   0   9   8
f   e   d   c   b   a

For information, write: Bedford/St. Martin's, 75 Arlington Street, Boston, MA 02116 (617-426-7440)

ISBN: 978-0-312-10201-2 (paperback)
ISBN 978-1-349-61481-3   ISBN 978-1-137-05068-7 (eBook)
DOI 10.1007/978-1-137-05068-7

Acknowledgments

Figure 1. KLEENEX® and DELSEY® Tissue Products are registered trademarks of and used with permission of the Kimberly-Clark Corporation.
Figure 3. Listerine is a trademark of and used with permission of the Warner-Lambert Company.
Figure 4. Used with permission of the Singer Sewing Company.
Figure 5. Used with permission of Star-Kist Foods, Inc.
Figure 6. Used with permission of Wearever/Mirro.
Figure 7. FORMFIT is a trademark of and used with permission of Jockey International, Inc.

Foreword

The Bedford Series in History and Culture is designed so that readers can study the past as historians do.

The historian's first task is finding the evidence. Documents, letters, memoirs, interviews, pictures, movies, novels, or poems can provide facts and clues. Then the historian questions and compares the sources. There is more to do than in a courtroom, for hearsay evidence is welcome, and the historian is usually looking for answers beyond act and motive. Different views of an event may be as important as a single verdict. How a story is told may yield as much information as what it says.

Along the way the historian seeks help from other historians and perhaps from specialists in other disciplines. Finally, it is time to write, to decide on an interpretation and how to arrange the evidence for readers.

Each book in this series contains an important historical document or group of documents, each document a witness from the past and open to interpretation in different ways. The documents are combined with some element of historical narrative — an introduction or a biographical essay, for example — that provides students with an analysis of the primary source material and important background information about the world in which it was produced.

Each book in the series focuses on a specific topic within a specific historical period. Each provides a basis for lively thought and discussion about several aspects of the topic and the historian's role. Each is short enough (and inexpensive enough) to be a reasonable one-week assignment in a college course. Whether as classroom or personal reading, each book in the series provides firsthand experience of the challenge — and fun — of discovering, re-creating, and interpreting the past.

Natalie Zemon Davis
Ernest R. May

Preface

Designed primarily for college courses in American history, American studies, women's studies, and popular culture, *Women's Magazines* brings together selections from mass-circulation magazines for women published between 1940 and 1960. By focusing on the World War II and postwar periods, this reader provides documents that participate in the debates about woman's role as homemaker and citizen, wife and worker, mother and consumer during two decades of enormous social change. With circulations numbering in the millions each month, the major women's magazines — through their editorial content, fiction, and advertising — served as advice manuals, guides to fashion and home decor, cookbooks, marriage counselors, and catalogs of goods and services. A sampling of their contents thus provides a resource for understanding how a pervasive segment of the popular press perceived and attempted to influence women's values, goals, and behavior.

The period that this volume addresses has been the subject of a great deal of recent reassessment by cultural historians, who have pointed out that American culture in the 1940s and 1950s was far more diverse economically, politically, and culturally than early television situation comedies would lead us to believe. On one level, the content of the women's magazines contributes to, rather than disturbs, stereotypical notions of the period: The overwhelming ideal in the magazines is the white, middle-class nuclear family. Yet a careful look at this material reveals glimpses of a world well beyond the tranquil suburbs. Advertisers' portrayals of African Americans almost exclusively as railroad porters and mammies expose the racial prejudice that was the target of the civil rights movement of the 1950s; the fiction published in the magazines frequently shows people coming perilously close to adultery or other illicit behavior; and the very fact that so many articles give advice on remedying troubled marriages suggests that Beaver Cleaver's family was not the norm in real life.

Nearly five years of research in the magazines themselves have convinced me of the centrality and continuing relevance of the following the-matic groups, into which the nonfiction articles are organized in this book: World War II, women and the workplace, marriage and motherhood, homemaking and consumerism, and fashion and beauty. It will quickly become obvious to readers that these are not mutually exclusive cate-gories. The material for this volume has been drawn from a dozen maga-zines, including *Ladies' Home Journal, Good Housekeeping, McCall's, Woman's Home Companion, Harper's Bazaar, Mademoiselle, Seventeen, Better Homes and Gardens, Redbook, Coronet,* and *Parents.* The first four periodicals in this list are represented most frequently in the volume for several reasons: They had the largest circulations during the period (although *Woman's Home Companion* ceased publication in 1957) and therefore had the greatest potential to both reflect and influence women's lives; because of their large subscriber and advertiser base, the magazines could afford to attract well-known writers to their pages; and they published articles on a wider variety of topics than did more specialized periodicals.

The book's final section, "Critiques of the Women's Magazines, 1946–1960," provides a different perspective on the content of the women's magazines of the period; it is composed of a group of articles critical of the women's magazines published in other periodicals during the period. Such critiques testify to the significance the magazines had assumed in midcentury American culture and can serve as starting points for discussion of the magazines' philosophies and contents.

Most of the authors whose names are not footnoted with their selec-tions were members of the editorial staffs of the magazines; a few were apparently freelance writers who did not achieve sufficient stature as writers to have biographical information included in standard reference works and computer databases. Although some of these authors pub-lished fiction and/or nonfiction in the magazines with some regularity, the fact that they published work primarily in mass-circulation periodi-cals has prevented their careers from being documented in the same way as have those of other writers. I take sole responsibility for errors or omis-sions in the documentation provided.

ACKNOWLEDGMENTS

I would like to thank the editorial staff at Bedford Books — especially Louise Townsend, Kate Sheehan Roach, Fred Courtright, and Stasia Zomkowski — for their assistance and their patience with what has turned out to be a much more complex project than any of us knew at the start. And my special thanks go to the staff in the bound periodicals department of the Ben West Public Library in Nashville, Tennessee, particularly Lawanda Smith, who made me welcome, assisted with the location of materials, and saw me through many an ordeal with copy machines and microfilm. Finally, the manuscript benefited from the thoughtful comments and constructive criticisms of Lois Banner at the University of Southern California, James Gilbert at the University of Maryland, College Park, Joanne Meyerowitz at the University of Cincinnati, Eva Moskowitz, and Ellen Schrecker at the National Humanities Center.

Nancy A. Walker

Contents

Foreword   iii

Preface   v

List of Illustrations   xiii

PART ONE
Introduction: Women's Magazines and Women's Roles   1

The Role of the Women's Magazines   1
Women and Society   11

PART TWO
The Documents   21

1. World War II   23

1. "What Do the Women of America Think about War?"
   Ladies' Home Journal, February 1940   24
2. Pearl S. Buck, "Women and War," *Ladies' Home Journal*,
   May 1940   26
3. Dorothy Dunbar Bromley, "Women Work for Their
   Country," *Woman's Home Companion*, December 1941   34
4. "How a Woman Should Wear a Uniform," *Good
   Housekeeping*, August 1942   35
5. "Meet the Berckmans: The Story of a Mother Working on
   Two Fronts," *Ladies' Home Journal*, October 1942   37
6. J. Edgar Hoover, "Mothers . . . Our Only Hope," *Woman's
   Home Companion*, January 1944   44

7. James Madison Wood, "Should We Draft Mothers?"
   Woman's Home Companion, January 1944   48
8. Alfred Toombs, "War Babies," *Woman's Home Companion,*
   April 1944   50
9. "When Your Soldier Comes Home," *Ladies' Home Journal,*
   October 1945   56

2. Women and the Workplace   **63**

10. Mrs. Franklin D. Roosevelt, "Women in Politics," *Good
    Housekeeping,* April 1940   65
11. "You Can't Have a Career and Be a Good Wife," *Ladies'
    Home Journal,* January 1944   71
12. Alice Hamilton, M.D., "Why I Am *against* the Equal Rights
    Amendment," *Ladies' Home Journal,* July 1945   76
13. Jennifer Colton, "Why I Quit Working," *Good Housekeeping,*
    September 1951   82
14. "Women in Flight," *Mademoiselle,* December 1952   86
15. "The Married Woman Goes Back to Work," *Woman's
    Home Companion,* October 1956   87

3. Marriage and Motherhood   **97**

16. Helen Welshimer, "My Husband Says — ," *Good
    Housekeeping,* February 1940   100
17. Wainwright Evans, "Are Good Mothers 'Unfaithful' Wives?"
    Better Homes and Gardens, July 1941   102
18. "What's on *Your* Mind?" *Redbook,* July 1945   106
19. Amram Scheinfeld, "Are American Moms a Menace?"
    Ladies' Home Journal, November 1945   108
20. "Are You Too Educated to Be a Mother?" *Ladies' Home
    Journal,* June 1946   114
21. "What Makes Wives Dissatisfied?" *Woman's Home
    Companion,* April 1947   116
22. Clifford R. Adams, "Making Marriage Work," *Ladies'
    Home Journal,* January 1948   121
23. Mrs. Dale Carnegie, "How to Help Your Husband Get
    Ahead," *Coronet,* January 1954   126
24. Paul H. Landis, "What Is 'Normal' Married Love?" *Coronet,*
    October 1957   136

25. Beryl Pfizer, "Six Rude Answers to One Rude Question,"
    McCall's, July 1960   141

4. Homemaking   145

26. Grace L. Pennock, "Starting from Scratch," *Ladies' Home
    Journal,* April 1940   146
27. Dorothy Canfield Fisher, "Housekeeping Need Not Be
    Dull," *Ladies' Home Journal,* October 1941   149
28. Dr. Carl P. Sherwin, "The Question-Box," *Good
    Housekeeping,* January 1944   152
29. M. F. K. Fisher, "The Lively Art of Eating," *Harper's
    Bazaar,* November 1944   156
30. Dorothy Thompson, "Occupation — Housewife," *Ladies'
    Home Journal,* March 1949   161
31. Helen W. Kendall, "Electric Mixers — Strong Right Arms,"
    Good Housekeeping, January 1950   165
32. Jane Whitbread and Vivian Cadden, "Granny's on the Pan,"
    Redbook, November 1951   166
33. Robert J. Knowlton, "Your Wife Has an Easy Racket!"
    American Magazine, November 1951   170
34. Paul Jones, "Is There a Plot against Women?" *Ladies'
    Home Journal,* July 1954   180
35. Selma Robinson, "103 Women Sound Off!" *McCall's,*
    February 1959   182
36. Joyce Lubold, "My Love Affair with the Washing-Machine
    Man," *McCall's,* June 1960   189

5. Fashion and Beauty   193

37. Ruth Anna Read, "Those Simple Little Exercises," *Good
    Housekeeping,* July 1940   196
38. "The Mood Has Changed," *Harper's Bazaar,* September
    1944   198
39. "Can You Date These Fashions?" *Harper's Bazaar,*
    October 1944   199
40. Sally Berry, "Do You Make These Beauty Blunders?"
    Good Housekeeping, April 1944   202
41. Elizabeth Pope, "What Is a 'Well-Dressed' Woman?"
    Redbook, July 1945   204

42. Janet Engel, "The Fattest Girl in the Class," *Seventeen*, January 1948   211

43. "The Lass with the Delicate Air," *Mademoiselle*, July 1949   215

44. "At My Age," *Harper's Bazaar*, December 1949   220

45. "The Line Forms Here," *Mademoiselle*, July 1952   221

46. Bernice Peck, "Accessory after the Body," *Mademoiselle*, October 1952   223

47. "Making Less of Yourself," *Harper's Bazaar*, September 1955   224

48. Elinor Goulding Smith, "How to Look Halfway Decent," *McCall's*, February 1959   225

6. Critiques of the Women's Magazines, 1946–1960   228

49. Elizabeth Bancroft Schlesinger, "The Women's Magazines," *New Republic*, March 1946   229

50. Ann Griffith, "The Magazines Women Read," *American Mercury*, March 1949   234

51. Marghanita Laski, "What Every Woman Knows by Now," *Atlantic Monthly*, May 1950   242

52. Mary McCarthy, "Up the Ladder from *Charm* to *Vogue*," *Reporter*, July 1950   247

53. Katharine M. Byrne, "Happy Little Wives and Mothers," *America*, January 1956   254

54. Joan Didion, "Marriage à la Mode," *National Review*, August 1960   258

APPENDICES

   Questions for Consideration   262
   Suggestions for Further Reading   263

Index   267

Illustrations

1. Delsey toilet paper advertisement from
   Woman's Home Companion   xiv

2. Cocomalt advertisement from *Ladies' Home Journal*, May
   1940   61

3. Listerine advertisement from *Good Housekeeping*,
   August 1949   98

4. Singer advertisement from *Woman's Home Companion*,
   December 1956   144

5. Star-Kist tuna advertisement from *American Home*,
   September 1950   155

6. Wear-Ever advertisement from *Ladies' Home Journal*,
   July 1946   179

7. Formfit advertisement from *Woman's Home Companion*,
   April 1942   194

8. "Can you date these fashions?" from *Harper's Bazaar*,
   October 1944   200

9. "Do you make these beauty blunders?" from *Good
   Housekeeping*, April 1944   203

10. "What is a 'Well-Dressed' woman?" from *Redbook*, July
    1945   208

Figure 1. Delsey toilet paper advertisement from *Woman's Home Companion.*

PART ONE

Introduction:
Women's Magazines and
Women's Roles

THE ROLE OF THE WOMEN'S MAGAZINES

Today information, entertainment, and advice come to us from a dizzying array of media, many of them electronic: film, television, video, and computer. Newspapers and magazines continue to play a role in influencing how we dress, conduct our relationships, cook, raise children, purchase products, spend our time, and plan for the future, but they must compete for our attention with the bolder, faster images and messages that arrive in our homes and workplaces at the click of a button. To understand the different role that magazines played in the lives of many women in the 1940s and 1950s, we must imagine a society very different from our own—one in which the only two technological links between the average American home and the larger culture were the telephone and the radio. The television set was not a common fixture until the late 1950s, and even then programming was extremely limited, by today's standards. Cable television service and the VCR were still in the future; the personal computer was undreamed of.

Magazines thus assumed more importance than they do today in helping to both shape and reflect the values, habits, and aspirations of American women and their families. Magazines designed primarily or exclusively for women readers had large circulations, mainly through subscription. In 1940, for example, *Ladies' Home Journal,* one of the oldest of these periodicals, claimed the largest circulation of any magazine

1

in the world, and during the 1940s and 1950s the leading women's magazines (which also included *Good Housekeeping, Woman's Home Companion, McCall's,* and *Redbook*) could boast of subscriber lists ranging from two to eight million. In terms of actual readership, the numbers were much higher: Issues were shared among family members and friends, and several of the magazines had regular features designed for the younger readers in the household. For example, *McCall's* had a Betsy McCall paper doll for young girls to cut out, and in the 1950s Abigail Van Buren answered questions for teenagers; *Ladies' Home Journal* offered advice to teenagers in its "Sub-Deb" column.

All of these magazines for women offered some mix of the following elements: fiction (primarily short stories, but sometimes serialized novels); poetry; articles on fashion and personal appearance; advice on household management—including cooking, cleaning, budgeting, child care, home decoration, and marital harmony; and features about or written by famous people. In varying degrees, the magazines ran articles on issues of wide cultural significance beyond the home, though always selected according to the editors' perceptions of women's roles and interests: for example, profiles of presidential candidates, articles on advancements in medicine, the advantages and disadvantages of television, and the American educational system. Some magazines included reviews of films and books, some had floor plans for houses, and some included sections on home repair or gardening. As important as this fiction and nonfiction content was, however, the most prominent feature in each issue of each magazine was advertisements, which occupied far more space than any other kind of material. In some issues of fashion magazines such as *Harper's Bazaar,* a reader had to flip through more than one hundred pages of advertising before even finding the table of contents for the issue. As much as the magazines were intended to entertain, enlighten, and advise women, they were overwhelmingly dedicated to selling products—from deodorants and canned soup to washing machines and girdles. Publishers kept subscription prices relatively low (*Good Housekeeping's* cover price in 1940 was 25 cents; in 1944, an issue of *Woman's Home Companion* cost 15 cents; the cost by subscription was lower than this) by attracting millions of dollars of advertising revenue; the volume of advertising meant that by the end of the 1940s, issues of some magazines were well over three hundred pages long.

The multipurpose "service" magazines were not the only kind of periodical designed for a primarily female readership in American publishing history, although they have been the longest-lived and, by the start of World War II, they vastly predominated in sales. Beginning in the nineteenth century, a number of periodicals were devoted to advocating women's rights, including but not limited to suffrage. The *Woman's Journal* existed in various forms from 1870 to 1931; others included the *Revolution*, *Woodhull and Claflin's Weekly*, the *Una*, and the *Woman's Advocate*. A few magazines were devoted to the woman who worked outside the home, ranging from the *Voice of Industry* in the mid-nineteenth century to the *National Business Woman* in the twentieth. The prevailing ideology that equated women with the domestic sphere, however, together with the technological developments such as vacuum cleaners and frozen foods that were marketed to women through advertising, ensured that such magazines as *Ladies' Home Journal* would attract the largest numbers of both readers and advertisers at midcentury.

In addition to the service magazines, several more specialized magazines attracted women readers during the 1940s and 1950s. *Harper's Bazaar* was the oldest of a handful of magazines whose primary emphasis was fashion. Begun as a weekly magazine in 1867, with the subtitle "A Repository of Fashion, Pleasure, and Instruction," it became a monthly in 1901 and appealed to the same economically upscale audience as did its chief competitor, *Vogue*. Readers were expected to have sophisticated tastes—not only in fashion, but also in food, reading, and travel—and plenty of money. During World War II, for example, articles on cooking in *Harper's Bazaar* reflected none of the deprivations due to food rationing that informed the more general-purpose magazines; readers were more likely to be informed by the French writer Colette about the heroism of European women than about making do without butter or new vacuum cleaners. *Mademoiselle,* marketed to a younger, more middle-class woman, began in 1935 as "the magazine for smart young women." The term "smart" referred not only to the dominant fashion emphasis of the magazine but also to its appeal to the female college student and college graduate. The ideal *Mademoiselle* reader of the period began reading the magazine as she chose a college to attend and a wardrobe to take with her, used it as a job-hunting guide when she graduated, and followed its advice when she furnished her first apartment. After her marriage (for the magazine assumed that the single state represented by its title was temporary), she quit work and turned to *Good Housekeeping* or *Ladies' Home Journal* for advice on how to make casseroles, make her marriage work, and make purchases for the home—unless, that is, she had come to depend on *Mademoiselle*'s impressive roster of contemporary authors who wrote both fiction and nonfiction for the magazine: Truman Capote, Carson McCullers, Russell Lynes, Jessamyn West, M. F. K. Fisher, W. H. Auden, Katherine Anne Porter, Jean Stafford.

Also appealing to younger readers interested in clothes, dating, and the culture of young adulthood were *Glamour* and *Seventeen*. Aimed at a somewhat less sophisticated readership, both had evolved from prior existences as Hollywood fan magazines—*Glamour* from *Glamour of Hollywood* in 1939 and *Seventeen* from *Stardom* in 1944—and both emphasized the concerns of the single young woman. Indeed, as its title suggests, *Seventeen* was for the teenage reader, whereas *Glamour* was for women aged eighteen to thirty-four. The evolution of two other magazines into magazines for women helps to illustrate the importance of the growing market of women readers after World War II. *Redbook* began in 1903 primarily to publish fiction. Well before 1940, each issue included several short stories, often a novella, and a complete novel, and during the 1940s the magazine published work by such well-known authors as Alec Waugh, Sherwood Anderson, Somerset Maugham, Philip Wylie, and Pearl S. Buck. By 1945, the proportion of the magazine devoted to fiction had shrunk considerably, replaced by general-interest nonfiction articles, and by 1951 *Redbook* described itself as "The Magazine for Young Adults." That the "young adult" was primarily female is made clear by ads for perfume and diapers; articles about childbirth, children's toys, and beauty; and increasingly formulaic fiction about relationships. The transformation of *Cosmopolitan,* which occurred later, was even more dramatic. Begun in the late nineteenth century as an intellectual magazine on a par with *Harper's* and *McClure's,* it featured articles on public affairs, science, travel, and education—at one point even attempting to sponsor a university by offering correspondence courses. In the early 1960s, its new editor, Helen Gurley Brown, changed it into a somewhat more daring rival of *Mademoiselle,* also aimed at the young single girl.

The contents of each magazine reflected its editor's judgment of what readers were interested in and wanted to know, from choosing a college or a winter coat to attracting and feeding a husband. While it would be impossible to know precisely what role any of these magazines played in the lives of American women during and after World War II, there are several important indications that they had a significant part in defining women's aspirations regarding work and family, appearance, health, and happiness. One indicator is the magazines' expanding readership: Their circulations increased from several thousand at the turn of the twentieth century to millions by midcentury, an increase achieved even as the number of different magazines aimed at the female reader increased from a handful to more than a dozen. Also, despite criticism to the contrary, the editors of women's magazines did not make choices about the contents of the magazines in a vacuum; indeed, the relationship between the editors and readers of many of the magazines was remarkably interactive, so that editors' decisions about regular features, special articles, and format were informed at least in part by expressed reader preferences. Some of the magazines regularly conducted polls of readers on selected topics; some invited questions from readers to be answered in print; all of the magazines received numerous letters from readers, and some printed a selection of these in each issue, creating a kind of forum in which readers spoke not only to the editorial staff of the magazine but also to one another. One editor, who spoke frequently to women's clubs, routinely asked the women to tell him why they bought magazines, and his favorite response was printed in the March 1942 issue of *Woman's Home Companion.* It reads in part: "We all want but one thing from our magazines: inspiration, . . . something that helps us make our lives better and richer with the beauty of living." While not all women would have found the magazines inspirational to this extent, the fact that their content addressed so many facets of women's lives makes them an important source of information about the cultural history of the twentieth century.

Indeed, the very existence of mass-circulation periodicals designed largely to instruct women in their appearance, duties, and values reveals fundamental differences between attitudes toward female and male gender roles. As Marjorie Ferguson points out in *Forever Feminine,* a study of British and American women's magazines, "there is no men's periodical press in the same generic sense that there is for women." A man typically has found more specialized publications directed at business, sports, or hobbies, not at "the totality of his masculinity, nor his male role as such." The implicit assumption, Ferguson continues, is that "a female sex which is at best unconfident, and at worst incompetent, 'needs' or 'wants' to be instructed, rehearsed or brought up to date on the arts and skills of femininity." Further, the heavy reliance of the magazines on advertisements for products, beginning at the turn of the century, adds the element of the woman-as-consumer, so that the magazines provided, in Ferguson's words, "a particular female world view of the desirable, the possible, and the purchasable."[1]

Women's magazines have long had their critics (see chapter 6). In 1939, in *America in Midpassage,* historians Charles A. Beard and Mary Ritter Beard wrote scathingly about what they perceived to be the nonintellectual content of the mass-circulation magazines, particularly their dominance by male editors:

> Owned and managed almost exclusively by men, as commercial enterprises, the journals "for women" automatically registered men's ideas of the audience to which the appeal was directed, and their enormous circulations implied that the managers correctly gauged the women to whom they supplied month by month fashion plates, fashion articles, society gossip, tepid fiction, bloodless sentimentality, Cinderellas, Fairy Princes, directions for the use of cosmetics, advertisements of the "allure."[2]

The Beards did see signs of a departure from what they termed "the deadly dominance of the commonplace" in the *Ladies' Home Journal* editorship of Bruce and Beatrice Blackmar Gould, who took over the magazine in 1935 and who attempted to introduce more public affairs content into the pages of the *Journal.* But David L. Cohn, in his 1943 book *Love in America,* was even more blunt: "An examination of a large number of women's magazines convinces me that their editors have a corrosive contempt for the intelligence of their readers, just as night-club owners have a similar contempt for the intelligence of their customers."[3]

But in fact the women's magazines of midcentury did not view it as their primary role to engage in debates about broad political or social issues. Instead they—especially those that had begun in the late nineteenth century—were direct descendants of *Godey's Lady's Book,* edited from 1837 to 1877 by Sarah Josepha Hale, who believed that women had talents and responsibilities quite different from—though no less important than—those of men. Hale also believed that women deserved training and support for their roles as homemakers, just as men were educated for their own professional lives, and she intended the *Lady's Book* to assist with their education *as women.* When twentieth-century editors spoke of the purposes and readers of their magazines, their remarks tended to echo Hale's. Herbert R. Mayes, who was editor of *Good Housekeeping* and then *McCall's* during the years 1939–69, reports in his autobiography, "It never occurred to us that advice on how to look and dress and cook better, how to make a home lovelier to live in, was less important than a polemic on the Christian ethic. The multimillion-circulation magazine was under no obligation, and isn't, to try to sway the populace with proposals for reforming society."[4] Bruce Gould, in *American Story,* indicates that it was what he and Beatrice remembered about their own families that prepared them to edit the *Journal:* "Beatrice was remembering her own dishwashing, bread-baking, poetry-loving mother. I recalled my family, straitened in money matters, never in intellectual aspiration or clear-cut moral attitude."[5] Gertrude Lane, who served as editor of *Woman's Home Companion* from 1919 to 1941, was as determined as Hale and the Goulds to regard her typical reader with a certain kind of respect:

> She is not the woman who wants to do *more* housework, but the woman who wants to do *less* housework so that she will have more time for other things. She is intelligent and clearheaded. I must tell her the truth. She is busy; I must not waste her time. She is forever seeking new ideas; I must keep her in touch with the best.[6]

Lane's concept of "the best" affected her selection of fiction to publish in the *Companion* more than it did an editorial emphasis on controversial issues; during her tenure as editor, the magazine published what we would call "serious" fiction by such authors as Willa Cather, Ellen Glasgow, and Sinclair Lewis.

If editors of the women's magazines assumed that their readers were intelligent, eager for information, anxious to perform their jobs well (whether in the home or outside it) and to look attractive while doing them, they also assumed—without ever saying so directly—that they were white and either middle or upper middle class or aspiring to be. Herbert Mayes refers to the assumed readership of *Good Housekeeping* as "middle Americans. Middlebrow. In every way middle."[7] From the editorial content to the images in advertising, there is scarcely a hint of any appeal to or recognition of the African American woman. The reasons for this are no doubt complex, a combination of unexamined cultural racism and concern for the business bottom line—advertisers and subscribers during the period would not have reacted favorably to articles about issues of primary concern to blacks or to pictures of black women using vacuum cleaners, unless, as was occasionally the case, they were obviously servants. *Journal* editor Bruce Gould relates an anecdote that provides chilling evidence of the economic power of racism: When, in its "How America Lives" series, the *Journal* profiled the family of a black Philadelphia physician and referred to his wife as "Mrs." rather than by her first name, circulation in the South dropped by two hundred thousand, and thousands of people sent the magazine letters of outrage.

It was not until *Essence* was launched in 1970 that there was a widely circulated magazine for African American women. (The Chicago-based *Tan* ran from 1952 to 1971.) *Ebony*, which was first published in 1944, was intended as the black counterpart to such magazines as *Life* and *Look*. Heavily dependent on photographic layouts, *Ebony* profiled prominent African Americans and featured articles supporting and celebrating the endeavors of blacks, men as well as women. Even in *Ebony*, especially during the 1940s, advertisements for products were often the same as those in magazines for white readers—complete with images of white people. Even when the product was one whose market would have been largely blacks, visual images reinforced the norm of whiteness. In an ad for a hair straightener, for example, the woman who is pictured to demonstrate its effectiveness is very light-skinned and has distinctly Caucasian features.

It may seem ironic that during the decade of the 1950s, when the civil rights movement was at its peak, the magazines for women seldom alluded to issues of race. The Supreme Court school desegregation decision was handed down in 1954; the Montgomery bus boycott began the following year; and in the fall of 1957, violent confrontation attended the effort to integrate Central High School in Little Rock, Arkansas. Almost none of this could have been learned by reading women's magazines of the period. Yet the magazines did not shy away entirely from sociopolitical issues. Evidence of cold war tensions between the United States and the Soviet Union, for example, was consistently present in the magazines in the years following World War II, especially in articles exploring the adequacy of the American educational system, but also in allusions to bomb shelters and civil defense plans. But there are important differences, primarily that whereas the threat of nuclear conflict with the Soviet Union could unify Americans in the face of a common enemy, racial integration was divisive. Furthermore, when cold war politicians invoked the "American way of life" as that which must be protected against the Soviet threat, they referred to precisely what the magazines existed to serve: the nuclear family, living in peace, able to take advantage of the capitalistic system to improve its standard of living. Appearing to take sides on the explosive issue of racial integration would doubtless have cost a magazine considerable advertising revenue, and it is important to keep in mind that the Beards were correct when they referred to the magazines as "commercial enterprises."

While the magazines by necessity tended to preserve the status quo in their editorial content and stances, they did not, as Betty Friedan claimed in her 1963 book *The Feminine Mystique,* consistently promote homemaking as the only path to female fulfillment. To be sure, important aspects of the magazines do convey this impression: The pages and pages of advertising depict women using electric mixers, washing machines, and various cleansers, but seldom driving a car or shopping for a business suit; and rare is an issue without an article on the problems of marriage and their solutions. But the magazines also regularly celebrated women's achievements outside the home, most obviously by publishing profiles of and interviews with well-known women, including film stars and other celebrities but also women who had made their mark in business, politics, and volunteer activities. In 1940, *Good Housekeeping* published a three-part series of articles called "Women in Politics" by Eleanor Roosevelt, and similar articles were not uncommon. *Mademoiselle* regularly ran articles on how to find employment in a number of fields, even though the repeated emphasis on women knowing how to type and take dictation suggested that they would be starting in secretarial positions. Somewhat less obvious, but equally important, is the fact that the magazines published both fiction and nonfiction by women, many of whom were professional writers. And it is easy to overlook the dozens of female staff writers for the magazines; these women may have been writing sober articles on how to make slipcovers or breathless reports on the latest Paris fashions, but they were being paid salaries to do so. In 1952, *Mademoiselle* published a follow-up study of 230 women who, as college students, had been "guest editors" of the magazine during the previous dozen years. More than half were married, and more than half of those who were married had jobs outside the home, about 40 percent of them as writers or editors. Although this represents a small fraction of thirty-year-old college graduates in 1952, it nonetheless held out to young readers a model of achievement beyond the cake mix.

And yet, embedded in the article *about* its guest editors is a reference to the instructions (elsewhere in the issue) for applying to *be* a guest editor, which serves as a reminder that the overwhelming role of the non-fiction portions of the women's magazines was to give advice on everything from choosing, buying, and wearing clothes to using an electric appliance. If advertising advised readers on which products to buy, the editorial content both told them how to use such products and implicitly promised that life would be better if they did so. In this sense, the articles themselves were a form of advertising, sometimes promoting categories of products (electric stoves, deodorants, new fabrics), but just as often mentioning brand names and thus endorsing particular products. Magazines devoted primarily to fashion, such as *Mademoiselle* and *Harper's Bazaar,* not only used brand names for clothing and cosmetics but often provided prices and the names of stores where the items could be purchased. Such magazines tended to use the imperative in the titles of fashion articles, suggesting that they were providing not just advice but orders: "Slip into Silk," "Take a Lesson," "Keep Your Hair On," "Make a Blouse." In the more general-purpose magazines, articles that promised advice tended to be headed with the phrase "How to . . . " or to ask a question of the reader: "Are You Too Educated to Be a Mother?" "Do You Make These Beauty Blunders?" "Are You Afraid of Childbirth?" "Are You Likely to Be a Happily Married Woman?" (In case the answer to this last question was no, an article in another magazine a few years later advised "How to Stay Married Though Unhappy.")

In their role as advisers to wives, mothers, homemakers, and to a lesser extent career women ("Want a New Job? Be Sure You Look the Part"), the magazines took on a function that we might assume had earlier been that of a young woman's mother. However, Sarah Hale's concern in the nineteenth century that the *Lady's Book* educate its readers in what came to be called "domestic science" suggests that a nostalgic vision of mothers teaching their daughters to cook, sew, budget, and dress is just that: nostalgia for a pattern that was never widespread or uniform. In the nineteenth century, the void was filled by authors and educators such as Lydia Maria Child, Catharine Beecher, and Sarah J. Hale, whose writings constituted motherly advice. By the end of the century, advice-giving became a community affair, with readers of magazines such as *Good Housekeeping* and *Ladies' Home Journal* contributing hints on household management for the benefit of other readers. The mid-twentieth century, however, was the era of the "expert," as psychologists, dieticians, marriage counselors, sociologists, and pediatricians (such as the famous Benjamin Spock) dispensed "scientific" advice in the pages of the magazines. Authors of countless articles (most, but not all, men) were duly credentialed as possessing advanced degrees and/or serving as heads of various "institutes." For the midcentury reader, there must have been a substantial difference between being advised by a fellow reader on how to remove berry stains from linen napkins and being instructed by a child psychologist on how to raise sons who were not "sissies"—not only was the fate of the son a far weightier matter than the fate of the napkin, but the person dispensing the advice was no longer a peer (to whom one could presumably offer the recipe for the blueberry muffins that stained the napkins), but someone possessing specialized knowledge to be conveyed from a position of authority. And the range of topics on which advice could be given was large indeed. A pair of articles in a 1940 issue of *Good Housekeeping* told readers how to read books and newspapers, advising in the latter instance to read only those columnists with whom one already agreed, to save "nervous wear and tear" and because "nobody was ever convinced by an argument, anyway."

In view of the close ties between advertising and editorial content, it is not surprising that the ads themselves often depicted an "expert" solving a woman's problem with a particular product. With the help of her son's teacher and the family doctor, a woman learns that Ralston cereal for breakfast will make her son a better student. Similarly, a woman whose son is "thin and nervous" because he does not eat well, when warned by her husband that "it's high time you *did* something about it," learns from a doctor that the solution is Ovaltine. Another woman, accused by her husband of serving boring salads, is hauled before the judge of "radio court" and admonished to vary her Wesson oil salad dressings; in the final panel of the ad the judge responds to her fervent "I don't know how to thank you" by intoning, "I'm glad you've learned your *Wesson.*" Sometimes the "experts" are other women, fellow home-makers who have already learned such "lessons" as that a Frigidaire electric range makes cooking "a lot of fun" or that Morton's salt doesn't

clump up embarrassingly in wet weather. Occasionally a product manufacturer offered advice that was wholly unrelated to the use of the product: A 1944 ad for Kleenex facial tissues instructed women on "what to
tell your husband if he objects to your getting a war-time job."
While it is impossible to know how most women responded to this
wealth of advice — or, indeed, to the magazines' total content — there are
indications that not all readers were grateful for it or intimidated by it. Evidence of resistance to various aspects of the magazines can be seen in
letters to the editors that most magazines regularly printed. Readers took
issue with the inclusion of certain subject matter, complained about the
fiction, and expressed resentment at being told how to conduct their
lives. *Redbook,* perhaps because it had originated as a magazine devoted
to literature rather than the household, was particularly open to expressions of opposition. A California reader wrote to *Redbook*'s "What's on *Your*
Mind?" column in 1945 to complain about the "advisers" who tell women
"how to hold a man, how to attract a man, how to please a man, how to
keep up his morale, and so on"; she requested (apparently to no avail)
articles telling men how to please women. And in a 1953 *Redbook* article,
two women authors created the hypothetical case of a woman driven
insane by all of the advice she was given about how to care for her baby.
The impression that these were not isolated reactions seems borne out
by the takeover in 1970 of the *Ladies' Home Journal* offices by one hundred
women representing several feminist groups. Claiming that the *Journal*
"creates frustrations which lead to depression and anger because women
cannot live up to what the magazine tells them they should," the women
demanded a regular column written by feminists and the ouster of editor
John Mack Carter.[8] The episode shows the influence of Betty Friedan's
attack on women's magazines in *The Feminine Mystique,* which had a profound effect on the women's movement of the 1960s and 1970s; it also
demonstrates the power that many people believed, rightly or wrongly, the
magazines had to affect women's sense of competence and self-worth.

WOMEN AND SOCIETY

The period between 1940 and 1960 was a transformative one for Americans in many ways. The United States participated victoriously in a world
war; the country's economy moved from Depression-era weakness to
postwar prosperity; the suburb became the ideal — and for many the
reality — of American life; developments in science and technology made
commonplace the automobile, frozen foods, television, and the bomb; the
civil rights movement was engaged and the seeds of the modern women's

movement were sown. When Friedan published *The Feminine Mystique* in 1963, she gave a name to one part of the complex ideology that women's magazines had provided for decades and attacked this ideology as leading to frustration rather than to fulfillment in women's lives. Yet so firmly entrenched was the "feminine mystique" that letters of outrage initially greeted the publication of Friedan's "The Fraud of Femininity" in *McCall's* in March 1963.

Statistics tell part of the story of two decades of change that was reflected in the pages of the major women's magazines. The percentage of Americans who were married increased from 60 percent in 1940 to 68 percent in 1960, and the number of families increased by 28 percent between 1947 and 1961. The size of the family increased as well: Between 1940 and 1945 alone, the birthrate increased from 19.4 to 24.5 per 1000 population. As though foreseeing this trend, the Goulds, upon assuming the editorship of *Ladies' Home Journal* in 1935, insisted that the architecture editor provide house plans with more than two bedrooms; house plans in the March 1949 issue of the *Journal* have three bedrooms, two of them labeled "boy" and "girl" (all master bedrooms in the feature are drawn with twin beds). The largest increase in the postwar birthrate was among the most highly educated women, and by 1956 one-quarter of urban, white college women were married while still in college.

While the magazines remained focused largely on women's role in the household, attitudes toward women's employment outside the home and patterns of such employment underwent dramatic shifts. A 1936 Gallup poll revealed that 82 percent of Americans opposed the paid employment of married women, and more than half of the then forty-eight states had laws that prohibited such employment in at least some circumstances. Shortly after the United States entered World War II, only 13 percent of Gallup respondents opposed women working, and the number of women in paid employment had increased by 60 percent by 1945, with eight million women entering the workforce. While the women's magazines generally supported women's *volunteer* activities in the war effort, there were exceptions, exemplified by the advertisement for Kleenex in the February 1944 issue of *Woman's Home Companion* that advised women how to insist on getting a wartime job. But for many women, these were to be precisely "wartime" jobs; within two years after the war ended, two million women had been displaced from their jobs by returning veterans.

Nonetheless, the relative prosperity that followed the war made possible a consumer culture for the first time since the 1920s, and there were scores of new products to be purchased. As late as 1941, more than one-third of American families lived below the poverty line, a fact that prompted articles such as "Feeding Five on $14 a Week" in the January 1940 issue of *Good Housekeeping*. During the 1950s, the number of families moving into the middle class (defined as having an income of $5,000 a year after taxes) increased by more than one million per year. During the five years after World War II, consumer spending increased by 60 percent and of this, the amount spent on household furnishings and appliances increased a dramatic 240 percent. In 1948, there were 500,000 television sets in American homes; four years later there were nineteen million. The migration to the suburbs was accompanied by an increase in automobile ownership. At the beginning of the 1950s, just under fifty million motor vehicles were registered; by the end of the decade there were nearly seventy-four million. In the June 1954 issue of *Ladies' Home Journal*, regular columnist Dorothy Thompson reported enthusiastically on her visit to one of the first shopping centers, in a suburb of Detroit; Thompson compared it to the central marketplaces of Greece and Europe and declared it "extremely practical" and "perfectly beautiful." (The "Making Marriage Work" feature in the same issue focuses on a woman whose problem is impulsive spending.)

The direct involvement of the women's magazines with the purchase and use of products long predated the 1940s. Several of them—*McCall's, Delineator,* and *Pictorial Review*—originated as ways to promote and sell paper dressmaking patterns in the late nineteenth century, when products ranging from carriages to corsets were also advertised in their pages. Even earlier, Sarah Hale, as editor of *Godey's Lady's Book,* had identified being a consumer as one of women's ways to exert power—even political power: During the Civil War she encouraged her readers to purchase American instead of British manufactured goods to support northern industry. Following World War II, women could exert similar political and economic force by purchasing goods that supported American industry and thus created jobs for returning veterans. The women who were the primary readers of the mid-twentieth-century magazines were not merely consumers, however; they were regarded as homemaking professionals who were assumed to be managing household budgets while on the alert for products with which to clothe, feed, and keep their families clean safely and effectively. Thus the magazines did not simply sell advertising space; they tested and recommended products—including, increasingly, electrical appliances—and instructed women in their use.

Nowhere is this link between advertising and editorial content clearer than in the history of *Good Housekeeping*. When it started publication in

1885, its subtitle was "A Family Journal Conducted in the Interests of the Higher Life of the Household." Much of the magazine's content was supplied by readers who submitted hints on cooking, cleaning, and decorating (in addition to poetry and fiction); and the editor, Clark W. Bryan, editorialized about product safety and effectiveness. After the Phelps Publishing Company took over the magazine in 1900 (at which point it had 250,000 subscribers), *Good Housekeeping* announced that it would refund the purchase price of any product advertised in its pages that proved to be unsatisfactory. By 1908, the Good Housekeeping Institute had its own testing facilities, and when William Randolph Hearst bought the magazine in 1911, he moved the institute to New York and installed as its director Dr. Harvey W. Wiley, who had drafted the legislation that created the Food and Drug Administration. Thus, while it might seem condescending—even absurd—to find in the January 1950 issue of *Good Housekeeping* an article about how to heat canned vegetables in a way that would preserve their nutrients, the magazine was merely continuing its long heritage of assisting the homemaker.

The fact remains that the women's magazines were—and are—businesses, and their dependence on advertising revenue for much of their income in turn meant that the editors took pains to avoid offending the manufacturers of soup, soap, deodorant, and washing machines. Bruce Gould recalls the strictures in force shortly after he and his wife became coeditors of *Ladies' Home Journal:* "They [advertisers] feared nude art (even if great), babies' bottoms, the mention of snakes in a story lest pregnant women miscarry. Words like contraception, coitus, even childbirth, were unthinkable. They would have swooned away at the mention of venereal disease in print."[9] The pressure of advertising on editorial content, which Gloria Steinem describes in vivid detail in her essay "Sex, Lies, and Advertising"[10]—coupled with the magazines' fierce competition with each other for subscribers—meant that the magazines remained reluctant to take on controversial subjects until those subjects had become fully accepted as part of a national conversation in which women readers might reasonably participate.

As responsive as they necessarily were to the demands of advertisers, the women's magazines also remained in touch with the opinions and needs of their readers long after the readers ceased contributing much of the editorial content. Most of the magazines conducted regular opinion polls of readers on a wide variety of topics, not all of them domestic. In its February 1940 issue, for example, *Ladies' Home Journal* reported in its "What Do the Women of America Think?" series—which ran from February 1938 to April 1940—that the overwhelming majority of respondents opposed American intervention in the war in Europe. *Woman's Home Companion* used what it termed "reader-editors," readers who wrote to the magazine to either give or seek advice about a particular domestic task and then were featured in subsequent articles concerning that task. And all of the magazines periodically featured question-and-answer columns that dealt with topics ranging from medicine to etiquette to personal appearance. Such efforts to involve readers in the content of the magazines (which also included the popular "makeover" features) not only continued long-standing tradition, they also allowed editors to stay in touch with the tastes and needs of their audience.

The extent to which the women's magazines responded to readers' existing tastes and needs and, conversely, the extent to which they shaped those tastes and needs cannot, of course, be precisely measured. Yet the magazines' essential functions as guidebooks and how-to manuals (how to dress, set a table, raise healthy children, stay on a budget, improve a marriage) meant that they inevitably presented a level of ideality to which women might aspire. That editors might have deliberately fostered such aspirations is evidenced in the Goulds' alterations in the *Journal's* fashion pages when they assumed its editorship in the mid-1930s. Concerned that the clothing then shown was "dowdy"—"average-priced average clothes, photographed on average women"—the editors decided to include "one beautiful spread of dream clothes" in each issue.[11] Even an article in the February 1959 issue of *McCall's* modestly titled "How to Look Halfway Decent" begins with the line "In order to be truly beautiful . . . . " Editors also could, and did, attempt to influence readers' thinking in more altruistic ways. Bruce and Beatrice Gould, for example, concerned about the relatively high maternal death rate in the United States in the late 1930s, commissioned a series of articles that advocated careful reviews of the circumstances surrounding any woman's death during childbearing—a series that angered the medical establishment but convinced readers of the need for medical accountability.

The most overt and far-reaching attempt to influence readers' values originated not with magazine publishers or editors but with the federal government. Immediately after the United States entered World War II, three agencies exerted considerable influence on the women's magazines to support the war effort—specifically, to encourage women to cope effectively with rationing and shortages, to do volunteer work, and, to a lesser extent, to enter the labor force. The War Advertising Council, the Writers' War Board, and the Magazine Bureau of the Office of War Information issued guidelines that affected the content of fiction and nonfiction features and the copy and illustrations of advertisements. From July

1942 until April 1945, the Magazine Bureau published the monthly *Magazine War Guide,* which was sent to hundreds of magazine editors. Instead of protesting the pressure to become conduits of pro-war propaganda, most editors considered it their patriotic duty to cooperate with these agencies. The results are obvious to anyone who leafs through a wartime issue of one of the magazines. An article in the December 1941 issue of *Ladies' Home Journal* tells women how to become volunteer workers through the Civilian Defense Volunteer Offices. *Good Housekeeping* for July 1942 instructs readers on its cover to "Buy U.S. War Bonds and Stamps," and an article inside provides advice on the proper care of vacuum cleaners—a product not manufactured during the war because of the military need for metal. Two months later in *Good Housekeeping* the war is credited with bringing about exciting new fashions through shortages of fabric: "We can be grateful to the government for providing us with a silhouette that is truly new." An advertisement for liquid pectin in the June 1944 issue of *Good Housekeeping* labels a woman making her own jelly a "portrait of a patriot."

Although, as Maureen Honey has pointed out, the *Magazine War Guide* and other propaganda sources did not explicitly pressure women to leave their war work for the kitchen after 1945, its advice to editors accomplished this purpose in more subtle ways. The magazines were asked to point out the need for workers in traditionally female fields, such as secretarial work and teaching, and to replace discussions of child care centers with articles on juvenile delinquency, "one of the social ills blamed on working mothers after the war which contributed to the postwar conservative reaction against working women."[12] As early as June 1944, a *Good Housekeeping* article featured designs for efficient "postwar kitchens"—the phrase providing a succinct description of the expectations for women's postwar work.

This abrupt shift in the magazines' presentation of women's roles—from active participation in a national effort to containment in a private kitchen—parallels William Graebner's characterization of the 1940s as two distinct half-decades, divided in 1945 by the death of Franklin Delano Roosevelt in April and the end of the war in the summer. In *The Age of Doubt,* Graebner describes the first half of the decade as public, nationalistic, and pragmatic, while the second half was private, familial, idealistic, and domestic. Yet despite differences in how women were expected to carry out their roles in each half of the 1940s, Graebner maintains that the roles themselves did not alter significantly: "Women spent the decade meeting the needs of men and capital; filling the factories as producers, then, after the war, soothing the fragile male ego, doing housework, and heading the family's department of consumer affairs."[13] And in spite of the fact that the war was a constant presence in the magazines between late 1941 and early 1946, in significant ways the role of homemakers as presented was unaltered throughout the decade. The tasks women are portrayed as doing during the war years may be described as somehow in the national interest, but they are still the domestic tasks of cooking, cleaning, and purchasing consumer goods. The August 1942 issue of *Good Housekeeping* offered advice in an article called "Shopping in Wartime for the Men of the House." In June 1944, *Woman's Home Companion* offered tips on inexpensive women's clothing in "Two Ways to Dress in Wartime." Cooking articles featured nonrationed foods or advised readers on ways to make scarce foods such as meat and butter go further, but they were nonetheless articles on cooking, and women were the ones expected to do it. No less famous a person than J. Edgar Hoover, director of the Federal Bureau of Investigation, writing in *Woman's Home Companion* in January 1944, insisted that women's "patriotic duty" was not on the "factory front" but on the "home front": "*There must be no absenteeism among mothers.*"

As the 1940s became the 1950s, the postwar emphases on consumerism, home ownership, and the nuclear family intensified, while at the same time so did the cold war and the Communist "witch-hunts." Indeed, the historian seeking the temper of the 1950s is confronted with contrasting realities. The cozy, bland, conformist culture depicted in television situation comedies such as *Leave It to Beaver* and *Ozzie and Harriet* and, superficially at least, in the women's magazines existed in idealized fashion against a backdrop of anxiety about nuclear war, violent confrontations over school integration, and the controversial Kinsey reports on Americans' sexual behavior. If the white, middle-class housewife in her prim apron — doing her laundry in high-heeled shoes — was one kind of ideal, Marilyn Monroe, who appeared in the first issue of *Playboy* magazine in 1953, was another. Even as television and the magazines promoted the happiness of the nuclear family in a tidy suburban home, institutions were established that would later be emblematic of movement away from the domestic haven: The first Holiday Inn opened in 1952, and McDonald's restaurants began to be franchised in 1955 (by 1960 there were 228 McDonald's in operation).

As a number of cultural historians in recent years have pointed out, the idealized family of the 1950s was not the culmination of long tradition; instead, it was a new phenomenon created by postwar America at least in part to celebrate democracy and capitalism in the face of the cold war threat of communism. In *The Way We Never Were,* Stephanie Coontz notes that "the emphasis on producing a whole world of satisfaction, amusement, and inventiveness within the nuclear family had no precedents";[14] and in *Homeward Bound,* Elaine Tyler May states that the 1950s family "was the first wholehearted effort to create a home that would fulfill virtually all its members' personal needs through an energized and expressive personal life."[15] The belief that "togetherness"—the theme of *McCall's* magazine beginning in 1954—was the best bulwark against an uncertain world may have reached its apotheosis in an August 1959 story in *Life* magazine about a couple who spent their two-week honeymoon in a bomb shelter.

In fact, by the end of the 1950s the image of the idealized suburban family and its material comforts had entered the realm of international politics. In 1959, then Vice President Richard Nixon visited Moscow for the opening of a trade fair and engaged in what became known as the "kitchen debate" with Soviet leader Nikita Khrushchev. As the two toured a model American house, Nixon angered Khrushchev by claiming that the Soviet superiority in rocket science was offset by the American middle-class standard of living. The democratic, as opposed to the Communist, way of life became crystallized in the leaders' debate as the freedom to choose one's own house and labor-saving devices. As Sonya Michel has pointed out, a public discourse that equated the family with democracy had gained force during World War II. At the White House Conference on Children in a Democracy, held in January 1940, President Franklin Roosevelt and others posited that the family was "the threshold of democracy . . . a school for democratic life."[16] The fact that the "teacher" in such a school was undeniably the mother placed a woman in a position of perilous responsibility: If she was too attentive to her (male) child's needs, she risked smothering him and making him "soft" rather than masculine; if she was not attentive enough, he might become a juvenile delinquent. To negotiate this dilemma she needed the advice of "experts," whose articles and question-and-answer columns became a regular feature of the women's magazines.

Despite their emphasis on women's domestic responsibilities, the magazines also sought to bring the larger world—at least certain parts of it—into the home by publishing book reviews, interviews with celebrities, and articles on education, life in other countries, foreign policy, and even psychology (usually popularized Freudianism). Both fiction and nonfiction by leading writers and public figures appeared in their pages, in part because the magazines were able to pay handsomely for such contributions. Sometimes these contributions were excerpts from forthcoming books, as when part of anthropologist Margaret Mead's *Male and Female* appeared in the September 1949 issue of *Ladies' Home Journal.* More often, authors were commissioned to write on topics the editors assumed would interest their readers. Thus, novelist Thomas Mann wrote for the August 1944 issue of *Good Housekeeping* about the Bible as a source of inspiration for his work, and Pearl S. Buck contrasted Chinese and American marriage customs in the September 1949 issue of the same magazine. During the same decade, John Steinbeck described to readers of the *Ladies' Home Journal* what life was like for American soldiers stationed in England, and Eleanor Roosevelt began her monthly question-and-answer column "If You Ask Me," which ran for most of the decade in the *Journal* and then ran in *McCall's* for most of the 1950s. In *Mademoiselle's* fifteenth birthday issue in 1950, the English poet W. H. Auden even mused about what he might have been like had he been female and an American as a reader of the first issue.

Reading a sampling of articles that appeared in mass-circulation magazines intended primarily for women readers between 1940 and 1960 opens a window on this period of history that no other single source can provide. As historians have come to recognize, the popular culture of a period—the books, films, music, and even the food and clothing that were experienced by large segments of the population—gives us insight into values, tastes, and desires that in turn affect political and economic decisions and trends. Magazines that were read by millions of women allow us to understand what society expected of them and, to a more limited degree, what women hoped for from life in American culture. Whether one views the magazines as having attempted to prescribe the ways women should be and behave or as reflecting what they valued, they offer a resource for understanding a crucial period of the American twentieth century.

NOTES

[1]Marjorie Ferguson, *Forever Feminine: Women's Magazines and the Cult of Femininity* (London: Heinemann, 1983), 2.
[2]Charles A. Beard and Mary R. Beard, *America in Midpassage* (New York: Macmillan, 1939), 741.
[3]David L. Cohn, *Love in America: An Informal Study of Manners and Morals in American Marriage* (New York: Simon and Schuster, 1943), 183.
[4]Herbert R. Mayes, *The Magazine Maze: A Prejudiced Perspective* (Garden City, N.Y.: Doubleday, 1980), 91.

[5]Bruce Gould and Beatrice Blackmar Gould, *American Story: Memories and Reflections of Bruce Gould and Beatrice Blackmar Gould* (New York: Harper and Row, 1968), 159.

[6]John Tebbel and Mary Ellen Zuckerman, *The Magazine in America, 1741–1990* (New York: Oxford University Press, 1991), 99.

[7]Mayes, *Magazine Maze,* 75.

[8]"Woman Power," *Newsweek,* March 30, 1970, 61.

[9]Gould and Gould, *American Story,* 168.

[10]Gloria Steinem, "Sex, Lies, and Advertising," *Moving beyond Words* (New York: Simon and Schuster, 1994), 130–68.

[11]Gould and Gould, *American Story,* 165–66.

[12]Maureen Honey, "Recruiting Women for War Work: OWI and the Magazine Industry during World War II," *Journal of American Culture* 3 (Spring 1980): 51.

[13]William Graebner, *The Age of Doubt: American Thought and Culture in the 1940s* (Boston: Twayne, 1991), 1–2.

[14]Stephanie Coontz, *The Way We Never Were: American Families and the Nostalgia Trap* (New York: Basic Books, 1992), 27.

[15]Elaine Tyler May, *Homeward Bound: American Families in the Cold War Era* (New York: Basic Books, 1988), 11.

[16]Sonya Michel, "American Women and the Discourse of the Democratic Family in World War II," in *Behind the Lines: Gender and the Two World Wars,* ed. Margaret Randolph Higgonet et al. (New Haven: Yale University Press, 1987), 155.

PART TWO

The Documents

1

World War II

Before the Japanese bombed Pearl Harbor on December 7, 1941—the event that brought the United States into the Second World War—American women's magazines were publishing articles about the war in Europe and the possibility of America's involvement in it. In fact, as a letter to the editor of *Redbook* in January 1940 makes clear, some readers wished that the magazines would devote less attention to war and return to their perceived functions of entertaining and providing advice. Articles about war before the attack on Pearl Harbor tended either to applaud the heroism of European women as they coped with wartime conditions or to propose that women's influence could keep America out of the war. Even as more and more American men were sent to military training camps, women were viewed as the peace-loving sex whose instinct was to protect their husbands, sons, and brothers from combat. As soon as the United States formally entered the war, however, the magazines adopted a pro-war effort stance that seldom wavered. Doubtless influenced in some measure by the *Magazine War Guide,* virtually all aspects of magazine content—including advertising, fiction, and editorial columns as well as nonfiction articles—instructed readers on ways to assist in the war effort: planting "victory gardens" to counteract food shortages, writing encouraging letters to absent husbands, and coping efficiently and cheerfully with product rationing and shortages.

Nor were women's war efforts to be limited to the household. Articles encouraged women to find volunteer work in hospitals and schools, and they presented admiring profiles of women who did so. In September 1944, an article in *Seventeen* asked young readers, "What Are You Doing about the War?" Readers were encouraged to buy war bonds, to report suspicious activities to the authorities, and to avoid purchasing black-market goods. Although the magazines seldom overtly encouraged women to look for *paid* employment outside the home during the war, they contain ample evidence that women were replacing men in factories and businesses. This acknowledgment most often took the form of concern about the children of working mothers: In the absence of adequate day care facilities, who would take care of them? Would maternal neglect cause an increase in juvenile delinquency? James Madison Wood, then president of a private women's college in Missouri, believed that women who were mothers *should* be drafted—to stay home and take care of their children. In general, the magazines dealt with the war and its effect on American women as a matter of how they could continue to fulfill their traditional roles of nurturer, consumer, and homemaker under unusual and stressful conditions, while at the same time acknowledging in myriad ways women's significant contributions to social stability during wartime.

1

"What Do the Women of America Think about War?"

Ladies' Home Journal, February 1940[1]

Do you think the United States should go to war to help England and France?

94 per cent said "No"
6 per cent said "Yes"

No flags flying, no bands playing, and no European crises—military or political—should stampede this nation into Europe's present conflict, according to the women of the United States. Almost unanimously—94 per cent to 6 per cent—they said "No" to the *Journal's* question, "Do you think the United States should go to war to help England and France?" Reasons they gave were realistic, echoing back to 1917. Said a 75-year-old mother of Marysville, Pennsylvania: "We had three sons in the Army during the World War and I know something of the horror of it."

[1]The series "What Do the Women of America Think?," based on polls of *Ladies' Home Journal* readers, began in February 1938 and ran until April 1940. Topics on which women were asked for their opinions included birth control, the uses of leisure time, the housewife's job, psychic phenomena, and whether Franklin Delano Roosevelt should be elected to a third term as president.

"What Do the Women of America Think About War?" *Ladies' Home Journal*, February 1940, 12.

Even should the Allies face defeat, the women are almost three to one against America entering the conflict. Asked, "Do you think the United States should go to war to help England and France if they are being defeated in war?" 74 per cent said "No" and only 26 per cent said "Yes."

Do you think the United States should go to war ONLY if this country is invaded?

77 per cent said "Yes"
23 per cent said "No"

Defending the home fires is another matter. To the above question, a 55-year-old saleswoman of St. Paul, Minnesota, said, "If they come over here I'll fight myself!" Seventy-seven per cent of the women agree that invasion is the only justification for war; only 23 per cent disagree. Mothers, naturally, feel most strongly about this. Among them, 79 per cent said "Yes" to this question, as compared with 73 per cent of women who have no children. Typical of opinion among the 23 per cent minority was the statement of the wife of a WPA[2] lunchroom worker of Asheville, North Carolina: "When the allies of the United States are in danger, we are in danger and ought to help." Most graphic—if most cynical—was the declaration of an unemployed waiter's wife, living in Manhattan. Said she, "All we got out of it the last time was a statue celebrating a soldier we don't even know the name of!"

---

[2]The Works Progress Administration (WPA) was a government agency established during the Depression to create jobs for thousands of unemployed Americans.

2

PEARL S. BUCK[1]

"Women and War"

Ladies' Home Journal, May 1940

When war was declared in Europe, the danger of our becoming involved seemed to many almost immediate. Memories of the last war, its sorrows and futile costs, pressed us hard; and yet the present Germany seemed necessarily as much an enemy as had been that earlier Germany in its threat to our democracy. We felt ourselves being compelled toward a decision we were not ready to make and had not, indeed, had time to make intelligently. For however one may plan in times of peace to take the permanent stand that war is not to be entered upon for any cause, each war, when it breaks, seems a particular war, different from any other in causes and possible effects. Those effects, especially, seem always more dire than the effects even of any previous war, and fear hurries us toward confused decision.

But there was a delay, and a delay valuable because the whole question of war, as it confronts us today, is a very different one from what it was months ago. We were given the great boon of time in which to gain a perspective upon war. In that time several things happened. There were some clear statements made by individuals upon the subject of war, and the general feeling of the people was also made clear. It seemed apparent—and is still, perhaps, apparent—that Americans do not want to engage in a war, at least to the extent of risking American lives. We are sympathetic in this European war, it is again clear, with England and France, and now especially with Finland, and we will help them all we can without taking the ultimate step, at least at present, of entering the war. The lifting of the embargo[2] was a distinct aid toward keeping us out of war—or, at any rate, toward postponing the date of our entry into it. In the same way the money proposed to be made available as a loan to Finland is of use to us.

[1] Pearl S. Buck (1892–1973), raised in China as the daughter of missionaries, won the Pulitzer Prize for *The Good Earth* (New York: John Day, 1931) and was awarded the Nobel Prize for literature in 1938.

[2] Buck may be referring to a 1940 modification of the Neutrality Acts of the 1930s that allowed the United States to sell munitions to its allies and thus participate in the war in an indirect way.

Pearl S. Buck, "Women and War," *Ladies' Home Journal*, May 1940, 18.

Deplorable as it is to think that there is to be made in our country the equipment of war whereby many men, women and children must die; nevertheless, in this one matter of keeping us at least temporarily out of war, the end of the embargo and loans to the countries with which we sympathize have been good, for they provide outlets for sympathies which run so strongly in us that, denied any way of expressing our sympathy, this energy might have forced us into other and even more dangerous activities.

And in the last place, we were given time in which to see how strange a war this is in Europe, and how confused are its issues. It is not merely a continuation of that war we once fought so hopefully to save democracy. We thought we had won that war—or some of us thought so, at any rate. There was a sort of victory, at least in armistice; but this war is being fought for something at once more deep than democracy, and less important than democracy, from the point of view of the rest of the world. That is, democracy is only one of the points involved, because democracy on the one side, as dictatorship on the other, is but a manifestation of a profounder difference between peoples, profounder and more instinctive than the difference between these ideologies.

The war is not between human ideas, in other words; it is between very human groups of people, who, unhappily, must live side by side, though they are so different that they should never be even on the same half of the world. In racial origin, in natural temperament and in the forms of education which have developed from those differences, such people as the English and the Germans are so different that oceans should divide them. Unfortunately this is not possible, and somehow or other they must learn to live side by side with their possessions overlapping, or else they must go on in these terrific periodic wars until one or the other of them is exterminated.

For peoples in nations have their antipathies exactly as individuals do. Every one of us knows what it is to have to live with or near someone whom we simply dislike, with whom no kindness or mutual forbearance can do more than keep strained peace, and we know that even such a peace is entirely dependent upon how far apart we can keep. When we must live as next-door neighbors, how difficult is peace! And yet, that peace would be no more assured if we involved in our hopeless antipathy other neighbors. Such involvement would merely mean enlarging the scene of strife and uncovering other lesser antipathies, which could remain dormant if undisturbed. This, on a small scale, is the present European war and what it would be if we were involved.

We have had time, too, to discover where we stand in relation to these causes for war which are so much deeper even than such ideologies as democracy and dictatorship. In this war of antipathies, where do we as Americans stand? Certainly not clearly on either side. Our profound belief in democracy lends its weight to make us turn from Germany and Russia and give our sympathy to those who fight against them. But when we go beyond this to the real cause of the war, we must find ourselves involved in both sides to a most confusing degree. We are, for instance, almost as much German in blood as we are English, and we have proved that, given the same environment, Germans make as good Americans as Englishmen do. We have, in short, too much German blood to make us complete allies of the English people and too much English and French blood to make us complete allies of the Germans, and if we enter this war it will have to be for reasons of our own and not for the same reasons that England and Germany are fighting each other. Their reasons we have not and can never have, because we are not a part of them even in the way that they are both a part of us.

For in our own fashion and upon our own soil we have already solved the problem which today in Europe they are trying to solve by still another war. English, French and German, we are living together here in peace and mutual respect because as individuals we have accepted something larger than any of us, and to a common larger group have contributed ourselves. I see this as the only solution for peoples as well as for individuals. It is a solution which, I believe, we can all hope is coming in spite of present war — the mutual, voluntary union of peoples into a federation which, recognizing no spoils of war, will give justice to victor and vanquished. England and France have begun already a sort of federation in their pooling of certain common interests in carrying on war, and such experience, even though it is among allies and not yet between enemies, is an important step in human education.

For this is the real way to found a federation — beginning not in some high arbitrary supercourt, but simply and practically in common needs for economy and convenience. The League of Nations as Woodrow Wilson conceived it was a vision, and it served the place of a vision. That is, it introduced to people an idealism far beyond them. It was doomed to fail because it was a vision and had no foundations in hard human experience. And yet, "where there is no vision the people perish." Having had a curtain of cloud lifted and having seen in heaven a state not upon this earth, though the cloud was dropped and the vision lost, it is not forgotten; and the idea of a federation among people which shall be for the good of all remains in many minds and is far more possible today than it would have been had Woodrow Wilson not dreamed, and even tried to make his dream work. That it failed does not mar its usefulness as a vision, therefore. That what people will make as a result of that vision must be something very different from the vision will not make theirs less good. In practical, slow ways certain unions are taking place between nations which may—I think will—carry over into foundations for a union which, however incomplete it is and should be for a long time to come, will, nevertheless, be the foundation for a world more stable and more secure than we human beings have ever known.

So much, then, for the value of delay. With every day of delay of our entry into war we are given precious time in which to reflect and to weigh the realities which are behind this war and to discover our relation to them. We may say now that we are too involved in them to be able to take clear issue on either side, and our real involvement with both sides makes it foolish to commit ourselves to a partial position with either side by going in to fight a little war of our own—say, for democracy—or even because our quickly warmed hearts flame at the injustice of huge Russia attacking little Finland.

Now I am not a pacifist, and anything I have to say about war is not, therefore, from the pacifist point of view. For I believe there are intelligent and necessary ways of using force. I believe there are times when force is the only way to settle a thing, either because it may be the quickest way to resolve an immediate situation which demands immediate resolution, or else because there are certain individuals who are of such a mentality that nothing except force, without admitting cruelty, can be used to control them. Thus I do not doubt that in certain primitive societies war is truly still inevitable, because the people in those societies are so simple and unreasoning that they give way to their instincts and cannot use their minds to encompass more than their own wishes and impulses. They are not able, that is, to reason beyond themselves and to consider the needs of others in relation to their own and to grasp the very practical truth that their own needs are better served in the end by reasonable ways of mutual understanding, if they can so achieve them.

War is inevitable among people who will not rationalize, but who will act only upon their own wants and angers. I see nothing to do about this except to allow such wars to proceed as a part of human education. Races and nations, we are not all in the same grade in the school of human experience. Some of us are only beginners in the lowest grades, and none of us are graduates. Only of one thing am I sure: in this school no grades can be skipped. We are each where we are, and none can help the others very much. We all learn as we are able to learn and we go at our own pace, some slowly, some quickly. We in the United States have gone quickly in this matter of learning to live together without war. Our marks are not perfect so long as we have the grave inequalities we have between races in our own country; but we are in a fairly high grade, nevertheless, when we consider our years. On the other hand, when we consider the favorable circumstances in which we have been able to learn our lesson in human living, our good fortune in being able to deal with individuals rather than with large solid groups such as there are in Europe, and with individuals already flexible and teachable or they would never have left those large solid groups, perhaps we have not done so well. Certainly the most conservative of English and French and Germans did not come to be Americans, and that may be the reason why those who stayed at home seem always to be fighting one another. Nevertheless, we as Americans have made a distinct success in mingling people from different countries and races, and we have no reason for joining in a war which, in its profoundest causes, is a war of human antipathies deeper than any ideologies.

Therefore, although I am not a pacifist, I see nothing in this present war, or in any war in history, which would make me proud to have a son go into it, or indeed make me do anything except try by every means in my power to keep him out of it. But this would be not from any pacifist ideas, but because wars, as I have studied them in history and experienced them many times in my own life, are either clashes between primitive peoples or else stupid outbreaks of antipathies which could be relieved in other ways. I can conceive of being very proud of having a son in an army made up by some future federation of peoples, an army whose sole purpose and duty would be to eliminate criminal individuals who were traitors to the federation, and who might gather about themselves, as gangsters do, other criminals and hold up a region or even a whole nation temporarily.

Such an army would be at the direction of no one nation, but at the command of all, under a head chosen by all, and would be supported by all. It would hold the same position in the world that a federal police body holds in a union of states, and would be called out only when local means of control had failed. It would be temporary, as we hope all police are temporary in our present society, though it would be folly to think we could live without them now. What we are doing internationally, of course, is a thing we would not dream of doing anywhere else: we are living internationally without a federal government and without a federal police, and consequently we are in a state of international anarchy, and we shall continue in that anarchy until we have the sense to realize how foolish it is. For there is no magic in mere numbers of people. A nation is nothing but a certain number of individuals, and the sum of their individual problems is the national problem, and the problems of the nations are the international problems. That is, a problem on an enlarged scale is the same old problem. Hence the difficulty England and France and Germany have in living side by side is the same difficulty you and I have in living next to someone whose race is not ours, whose language is not ours and whose ways are different from those we know and like. As individuals we have, however, learned to refrain from setting one another's homes on fire, and have been helped in this by the policeman on the corner, whom the community pays for just that purpose.

Well, what has all this to do with women? Simply this: that I believe the whole question of war and peace is a woman's question, and we can decide it as we will. If the thirty-seven million women of the United States should will not to go to war on a particular occasion, there would be no war. Each day of delay in deciding our share in the European War has, therefore, been of inestimable value to us as women, because it has given us time to think for ourselves—and if not to think, since not all of us are given to thought, at least to feel about this war, and to come to wonder if it is ever going to be a war worth our fighting. I have tried to suggest in what I have said that this war in Europe is really not our war and its causes have nothing to do with us.

And yet, there is the matter of sympathy. If our sympathy lies with one side as against the other, that is reasonable enough and only human. What shall we do when with all our hearts we desire to help, say, Finland against Russia? Are we to stand watching and doing nothing while innocent weak people are so attacked? Our quick blood, our aroused emotions say no—not Finland, at least! For it is true that we have let Japan go on attacking China for some years, we let Italy attack Ethiopia—surely as unjust an attack as Russia's upon Finland, and more unreasonable, for Finland did once belong to Russia for a long time. And we let some strange, wicked things go on in Spain without thinking about taking up guns.

Sympathy, therefore, is not a thing of pure justice. It is spurred on in the case of Finland by our hearty dislike of Communism, by our kindly feeling toward a little country honest enough, as we say, to pay her debts, and as much as anything by our real anger at a big something, man, dog or country, attacking a little one. All our hottest American feelings are ablaze when we see an underdog getting the worst of a battle, and naturally, for we are, as a nation, a haven for underdogs, and in us all memory quickens sympathy.

Of course in that word "sympathy" lies a great and dangerous power, dangerous because sympathy has seldom anything to do with reason. And more dangerous because sympathy is an instinct in which women are strong, a good instinct when it is combined with understanding and an evil one when it is without understanding. For sympathy, like other instincts of love, has its limit, and beyond it there is merciless cruelty. We sympathize with one's suffering and therefore we hate another who, we think, caused the suffering.

Sympathy clouds the understanding more often than it combines with it to work real good. It makes us yield to impetuous punishments and retaliations. It makes us consider ourselves as avenging gods, with the right to judge and to punish, even by death. The sinking of the Lusitania in the last war, for example, whipped American sympathy into such anger against Germany that Americans demanded war, and I do not for one moment doubt that the sympathy which did it was 90 per cent feminine—women in their homes saying to their men, "Are we going to stand this sort of thing? Think of those poor mothers and their children—we've got to do something about it!" And men, listening, went out and echoed grimly, "We've got to do something about it," and they were the men who went out to fight, and thousands more mothers lost their children.

If those women had been able to use understanding with their sympathy, and to have made their sympathy go beyond the immediate moment, they would have said, "Wait—this is horrible, but there is something more horrible: war, and thousands more of our sons, and the sons of other women in other countries, dead! There is, because there must be, another way." It is that wider sympathy, that sympathy discerning and intelligent enough to go beyond the immediate case, the present moment, which we who are women need to develop. Sympathy is genuine only when it is the result of intelligent feeling and understanding. When it is merely a nervous energy, born of prejudices and angers, it is as dangerous as dynamite exploding in the wrong place.

Do I mean to say, some woman asks, that women can prevent war? Well, at any rate, I mean to say this: I have never been in any country where the women were not able to do whatever they wanted to do if they cared enough about it to begin it and to carry it through to their determined end. Women do not care about many things, I confess, but I keep hoping they care enough about war not to have it. At any rate, I make the

flat statement—which I believe to be true, though I cannot prove it—
that whether the United States goes to war or not depends more upon
the women and whether they are willing to let their men go to war than
upon any other one thing. We women hold public opinion in our hands
partly because we are first-rate purveyors of it through the relay method,
and partly because we create it in our own homes. If we are determined
to discover all the reasons why we should not go to war, and believe what
we discover, and tell our men there is no war worth fighting in this day
and age, there would not be any war.

Of course our difficulties will be with ourselves. I do not mean among
ourselves; for, all stale jokes and commonplace quips to the contrary, my
observation is that women work together extraordinarily well when they
have something to work for. And I cannot imagine anything more worth
working for than the prevention of the sort of wars we have nowadays, by
women organized to see that every woman behind the shut doors of her
home is given enlightenment wherewith to control her instincts. For a
woman's real enemies are her own instincts. She lives much alone, and her
contacts are mainly with others like herself. And in loneliness the instincts
grow strong and the understanding does not keep pace with them.

I have touched upon the instinct of sympathy, which is a peculiar force
in women for good or evil, depending upon the understanding which
accompanies it. I shall touch on one more—it is our instinct for home. I
speak of it because, perhaps more than sympathy, it is the instinct which
leads women first to tolerate and then to support the idea of the necessity of war and then to idealize it and glorify it to the point of sacrifice of
the ones we love best, in whose death we ourselve are doomed to die.
What is the instinct in us stronger than the love even of an individual? It
is the instinct of the preservation of home. I do not mean a particular
house in a particular place. I mean home, the breeding place; home, the
sort of environment to which we are accustomed; home, the kind of place
we want in which to rear our children. When this home is threatened, we
send our men out to fight to save it. It is an instinct so deep, so secret, its
roots so entwined and entangled among other roots in our subconscious,
that we do not recognize it. But it is all the stronger for that. We all know
that women are the homekeepers, the conservatives of the race, the ones
to preserve that which has always been. This conservatism has its deep
roots in our primal function of breeding.

The wise woman, therefore, knowing the power of life and death she
holds in her hand, will ponder these things and make up her mind not in
the immediate moment, not on quick sympathy, not on the hidden instinct

for security at all costs. No, the wise woman——But there are not many wise women. None of us has wisdom enough for the great responsibility of deciding whether or not to send our men out to war.

Yet to know this at least is the beginning of wisdom: that only when we come to the initial place of the rejection of war as a practical means to any end can we set ourselves to discovering what is the practical solution to our fundamental human problem of living together with all our differences. Rejection of war as a solution to any human problem is not to be made on sentimental grounds; nor even, at least to my mind, on the high grounds of religion or pacifistic principles. War is to be rejected for two simple, practical and very well-known reasons. First, it is always too costly. There has never yet been a war which did not cost far more than it was worth to anybody. Second, war accomplishes none of the aims for which we wage it. It merely deepens antipathies and aggravates differences, and each outbreak of war postpones true and lasting peace still more indefinitely into the future.

The rational rejection of war, therefore, on completely practical grounds, is the first step to peace. It must be an individual rejection, made by individuals enough before it can become a group rejection, and, I believe, made first by women as individuals; because what women, in the enormous, secret power of the home, first will to have come about, is what comes about in the vast life of our great common human family.

3

DOROTHY DUNBAR BROMLEY

"Women Work for Their Country"

Woman's Home Companion, December 1941

One of the pleasantest ways to be patriotic is to go in for active sports or at least take setting-up exercises. To encourage physical fitness former tennis champion Miss Alice Marble has been appointed National Director of Physical Training for Women.

Dorothy Dunbar Bromley, "Women Work for Their Country," *Woman's Home Companion,* December 1941, 82.

To see that all our children have the right food and enough of it is about the biggest job we can do for America. As in this Michigan school women in many states serve hot lunches supplied by Surplus Marketing Administration. Perhaps you think you could drive a truck or an ambulance but to be really useful you need also to know how to repair it. These housewives are taking a course in motor mechanics at a vocational school in Syracuse, New York.

Since communications are among the matters vital to defense you might equip yourself for emergency service by learning to run a switchboard as this Seattle Junior Leaguer is doing.

You needn't pilot an airplane, nurse in the army or go into a munitions factory to be of help. There are any number of important though less dramatic things to be done. Volunteer as an air raid warden, drive soldiers and their families to and from camps, entertain at the service clubs and at home. At Travelers' Aid booths in a hundred cities trained volunteers hand out information to boys on week-end leave.

Take a home-nursing course or be a gray lady—a hospital assistant. Study nutrition at a Red Cross class and keep your family well, or study canteen work for larger-scale feeding. If you are qualified teach instead of learn. And when spring comes plant a vegetable garden.

Want to help but don't know where to go? Send 10 cents for What Can I Do?—our booklet on national defense. Address Woman's Home Companion, *Service Bureau, 250 Park Avenue, New York City.*

4

"How a Woman Should Wear a Uniform"

Good Housekeeping, August 1942

The Editors of Good Housekeeping, *in a response to a request for help, present these several rules.*

If men don't like women in uniform—and most of them don't—it's up to us to make them change their minds. After all, their scathing crit-

"How a Woman Should Wear a Uniform," *Good Housekeeping*, August 1942, 21.

icisms are not of the costume itself. Most of them admit it is often becoming. What they don't like is the way we wear it, when we wear it, and where. In these matters, the distinction between right and wrong is clear-cut. There is no hazy no-man's-land of doubt. Either one should or one shouldn't. So here, to guide you, is our little list. Shun the Don'ts, observe the Do's, and you will inspire as much approval in uniform as out of it.

Don't

Don't cut your hair very short. A mannish effect is the last thing you want.

Don't wear a long, flowing hairdo. Whether it's glamour bob, sheepdog, spaniel, or plain hound style, dripping hair doesn't look right with a uniform.

Don't have a shiny, undressed face.

Don't lean over backward to look serviceable. The result is too hard on your public.

Don't wear long nails; they are not practical. Don't paint them luridly.

Don't wear frivolous shoes with high heels, open toes, open heels.

Don't wear jewelry, except watch and wedding ring. No flowers, either.

Don't wear a uniform in a night club, cocktail lounge, or at a bar.

Don't wear it in the evening unless you are on assignment. Don't wear it on a date, especially if he is a service man. He sees enough of uniforms.

Don't smoke or drink in public while in uniform. Don't swagger or stride along in masculine fashion.

Don't assume a self-righteous air, as if you were doing more for your country than other women are.

Don't wear a uniform unless it has real significance and you are fully qualified to wear it.

Do

Do choose a feminine coiffure; but pick something simple, easy to manage.

Do cut your hair shorter, give it an upward curl. Or keep your hanks of hair if you want to; but pin them up or wear a net when you're in uniform.

Do look finished as well as well-scrubbed. Be conservative; but use any make-up you require to get the fresh, pretty, mind-on-the-job appearance you should have.

Do keep your nails rounded to a moderate length. Lacquer conservatively.

Do wear sturdy, low-heeled walking shoes, comfortable and appropriate.

Do think before you add a gadget, "Is this really useful?"

Do make it a personal, inflexible rule never to wear a uniform except when on duty. Do change to civvies the minute the day's work is done. Wear your uniform anywhere in direct line of duty—but respect its purpose.

Do wear your uniform modestly, simply, and unself-consciously. It is a working costume, a sign of service, and should be worn with the quiet air of a woman who, like anyone else, is sincerely trying to do her best.

Do earn your uniform. (One of the largest organizations requires fifty hours of service before permitting its members to buy a uniform.)

5

"Meet the Berckmans: The Story of a Mother Working on Two Fronts"

Ladies' Home Journal, October 1942[1]

96.77% of American families have incomes of less than $4500 a year.

"Women workers in ordnance will have a large share in determining whether we win or lose this war," according to Maj. Gen. Levin H. Campbell, Jr., head of America's $30,000,000,000 Armament Production Program. "Within the next year, 90 per cent of the workers getting out guns,

---

[1]This article was part of the *Journal*'s "How America Lives" series, which began in February 1940. In 1953 the series was retitled "How Young America Lives," and by 1961 the title reflected a narrowing of focus to "How America Spends Its Money." The original title and focus were restored in 1962, and the series ended in 1963.

"Meet the Berckmans: The Story of a Mother Working on Two Fronts," *Ladies' Home Journal*, October 1942, 95.

tanks, ammunition and optical instruments will be women. We're glad the girls are with us. They're going to help us win!" At present, however, not all war plants employ girls. Because of the acute housing situation, Washington wants women to stay put, find jobs near home if possible. Housewives most likely to replace men needed for war: Women under 45 with no children under 16.

Soon after last New Year's the boys working in the North Armory of the famous Colt's Patent Fire Arms Manufacturing Company in Hartford, Connecticut, got a new benchmate—a husky, firm-handed individual wearing brown gabardine work pants and trained in handling small metal parts by five years in typewriter factories. The only remarkable thing about that was the new hand's being a woman—a big, blue-eyed, fair-haired, lilting-voiced Irishwoman with the free stride of a Celtic goddess. Never before had Colt's allowed women to touch manufacture of their revolvers, pistols and machine guns. But Pearl Harbor made a quick difference, and before spring was over more than 10 per cent of the thousands of workers in the plants of the Colt Company were women, with Mary Godfrey Berckman one of the first hired.

Eight hours a day, seven days a week, hair coifed in a bright handkerchief against accidents in moving machinery, feet planted on a wooden box, this devoted wife, mother of four, is steadily putting finishing touches on the beautifully machined cylinders that hold the bullets in .38-caliber revolvers. Thanks to her quick laugh and Irish tongue, she and the boys get along the finest way on distinctly first-name terms. Her foreman, who was badly worried when they moved women in on him, now admits that he finds them both steadier and more adaptable than men.

That description certainly fits Mary Berckman. Six years ago, when heart trouble put her husband out of the running for a long time, she philosophically turned to and got a full-time job so she could bat for him as breadwinner. In exactly the same spirit of doing whatever an emergency calls for, she is now helping give the nation what it needs to win through, getting man's pay for man-sized work.

Dark, quiet, lanky, crag-nosed Fred Berckman has just as much philosophical realism in his make-up. When illness struck him, he displayed no male contempt for pitching in on anything he could do: while Mary brought in the cash from her shop job in the Royal Typewriter plant a hundred yards back of the Berckmans' house, Fred cooked, washed, ironed, cleaned and bossed the four youngsters with the same efficiency and care he had formerly displayed in grooming fairways for the Hartford Golf

Club. Over a year ago the doctor pronounced him back in shape and he went to work on the oil-cleaning centrifuges at the Royal plant, where he is doing all right. But he did not abandon his share of the household even then. Since he has Saturdays and Sundays off and Mary has not, Fred spends Saturday morning running the family underwear through the washing machine and ironing it afterward—he is a good ironer as well as a good plain cook—and, when she gets home Sundays, usually round four o'clock, he has a first-rate Sunday dinner waiting for her. Every school day he comes home for lunch to meet the youngsters, who have set the table against his arrival, and fixes them hot soup and sandwiches. Without his co-operation, they would have to have some kind of maid.

Mary feels that being a working mother can be a good thing, provided you have a husband sporting enough and youngsters smart enough to permit it at all. "It's good for a woman to get out and do something away from her family," she says. "You get irritable sticking with housework, especially with a lot of children. I know women who've gone so stale raising a family and polishing furniture they won't give you a civil answer to a civil question. If you can't stay home brooding over what's wrong with the kids, it works out better. You go to your job, meet a different crowd, get your mind right off it. Then things are much easier when you do get home."

Complete lack of days off is a strain, of course, but Mary can take it. The whistle blows at three, she is home by four and fixing chocolate malted milks for the youngsters' mid-afternoon refueling. Then she sheds her work slacks—they made her feel very queer at first, but they're second nature now—and lies down to doze until Fred gets back. Her nine months at Colt's have certainly harmed her not at all; her eyes are as sparkly, her skin as clear and her step as light as when she first landed in America from County Mayo at the age of nineteen.

Between Fred's $27-a-week average, $2 more than he used to make driving tractors for the golf club, and Mary's $54—85 cents an hour, time and a half Saturdays, double time Sundays—the Berckmans' current income is far and away the best they ever knew. Close to a dollar a day is going into War Bonds and Stamps. When back debts are paid off—just as great a contribution to the fight against inflation, says Leon Henderson—even more will go the same route. Since living costs are sharply up, too, however, the Berckmans are not yet living the life of Riley, for all their being three quarters Irish. With prices at local stores higher than they used to be and Mary no longer around to do the baking and supervise thrifty use of leftovers, Fred's pay just about covers food. Even though they have had no car since Fred was taken sick and the youngsters' parochial school costs nothing, the money still rolls out pretty fast. Fred likes his present job, but it hasn't the appeal of the old days when he could let in the clutch on a big Fordson tractor and feel her forward surge all the way up his spine. "I wore out two of those big fellows under me in the seventeen years I worked at the club," he says proudly. His father, a mason, of German-American parents, died when the boy was small in Rutherford, New Jersey. His mother, a New York Irish girl whose father was still coachman for a wealthy New York family, moved herself and four children in with her parents and got herself a saleswoman's job at Bloomingdale's department store in New York. Fred was reared in the patriarchal quarters that Grandpa Cottingham had over the coach house back of the big mansion on East Fifty-seventh Street.

A photograph of grandpa in professional silk hat, on the box of an elegant equipage, is conspicuous now in the Berckman's living room. The old gentleman knew boys in definitely old-fashioned terms—young Fred was acquainted with both the way a horsewhip stings when applied to small and erring legs and the delights of riding beside grandpa when the horses were taken downtown to be shod. When the dispute between Fred and his mother over the suitability of Lord Fauntleroy[2] curls became acute, grandpa solved it out of hand by taking his rebellious grandson into the coach house and going to work on him with the horse clippers.

After finishing eighth grade at the Cathedral School back of St. Patrick's, near by, Fred delivered parcels for a costumer at $3 a week, then did odd jobs at a primitive New York garage, then, when his mother married a Hartford gardener and moved the family to her new home, learned gardening from his stepfather on a Hartford estate. A chauffeur's post, a stretch with the Army transport service in 1917–18, saw him into the golf-club job and marriage. A plain, hard-working, brilliantly honest and superbly self-respecting man, every inch of his six feet the McCoy.

When introducing him to Mary Godfrey at an Irish ball in Hartford, a mutual friend carefully transformed Fred's German surname into "Murphy" to make sure the pair would get on well. They did—the only thing they have disagreed about yet is whether steaks should be broiled rare or well done. Mary liked Mr. Murphy better and better, even after she learned he was really Mr. Berckman. She taught him to perform an Irish

---

[2] Little Lord Fauntleroy, the title character in a children's story by Frances Hodgson Burnett, dressed in frilly clothing and wore his hair in blond curls.

jig to go with her own light-footed skill in that line — at thirty-eight, with plenty of meat on her slender bones, she can still foot it with the liveliest when she and Fred have their one big evening of the year at an Irish dance on St. Patrick's Day. He was a good Catholic, had a steady job, was quiet. "My friends said, 'He never says anything — how did you ever land him?' and I said, 'Ah, it was me did most of the talking!' " There were no misgivings at all when they were married at St. Patrick's Church in Farmington, Connecticut, fourteen years ago.

Poultry and cows, pigs, haying and potato digging on the twenty green Godfrey acres in Ireland, handed down from father to son for uncounted generations, had already given Mary a fine ability to turn her hand to anything when she "went into service" in America. As a nursemaid in well-to-do Hartford families, she made $80 a month and board, keeping other people's youngsters walking chalk and delighting them with the bubbling run of her busy Irish voice and her gorgeous head-tossing laugh, as healthy as sunshine for children.

She would have no objection at all to her own daughters' doing such work: "I couldn't be in service now myself," she says. "A house of your own spoils you for other people's. But, for a smart girl that can choose her place, it's a good life and grand way to save money." Catherine, her perky twelve-year-old eldest, has a mildly glamour-smitten ambition to become an air hostess, and Mary hopes that some one of the three girls may make a teacher. But she would also like to see some of them in one of those new professional nurse-maid training schools.

As for Junie, her youngest, a sturdy, frantically freckled and utterly Irish-featured boy of seven, she would give her right hand to see him become a priest. A good seminary education for him is the principal use to which she hopes to put those accumulating War Bonds. So far, being principally absorbed in the navigation of toy battleships and games of ball with daddy in the back yard, Junie hasn't troubled his bulletlike towhead much about that.

He will certainly get devout rearing in this good Catholic household, with holy pictures in every room and, on the old sewing machine in the dining room, a pretty little blue-and-white crepe-paper shrine containing a dainty white-plaster Virgin wearing a gilt-paper crown that Frederica made for her. All during Lent the children saved up the odd pennies they acquire in place of regular allowances, denying themselves candy or ice-cream cones, and then, after Easter, collectively splurged the hoard taking daddy and mommy to a restaurant dinner and the movies.

For all her seven-day-a-week schedule, Mary never misses Mass — she is up at five Sunday morning to go along to church with the fellow employee, also Irish and Catholic, who drives her to work on a share-expense basis. The children never miss either, trooping off together, often with daddy, later in the morning, each with a nickel for the offering safe in the right fist.

Mary thanks her stars that the three girls got her own fair and naturally wavy hair while Junie got a straight-haired cowlick—to make sure of that she kept cutting his baby fuzz as soon as it was long enough to take scissors. All these three little girls are as articulate, resilient and chipper without being fresh as any mother could wish, getting their own breakfasts on school mornings, making their beds, pitching in to help finish the dinner that Mary has started before her after-work rest period, wonderfully combining co-operation with mischief. Unself-consciously they run on and on about how daddy exiled their kitten when, while he was doing the housework, it made a mess in each room in succession; how they brought down maternal wrath on themselves—when much smaller, of course—by giving one another shingled haircuts in private after seeing the hairdresser working on their mother; accurate reflections of the present relative prestige among Hartford's small fry of Ella Cinders[3] and Superman; crisp comment on Frederica's precocious ability to bake a good cake at the age of eleven and Eileen's nine-year-old penchant for wearing her best clothes out of season.

Although their school grades are nothing outstanding, their special quality of human rightness and brightness impresses even the veteran sisters at the parochial school right behind the Gothic parish church of Our Lady of Sorrows: "Mrs. Berckman," they always say, "you have nothing to worry about in these four of yours." Many a worried mother would regard their spontaneous well-being as an utter mystery, considering that for the past five years their mother has been on a factory job at least five days a week. Even when mild childhood diseases have come along, they could juggle their own medicines during the never-more-than-four-hour period when no adult came in, an older sister staying home to take care of Junie when necessary. Good will and serenity, coupled with expert co-operation from Fred, and a proper dash of Irish fire when occasion demands, have been the successful recipe.

The Berckmans' neighborhood—two blocks between a dead end

---

[3] *Ella Cinders* was a comic strip created by Charles Plumb and Bill Conselman; it ran from 1925 to 1961. A film version, a romantic comedy that spoofed the Cinderella story, was made in 1926. It starred Colleen Moore and Lloyd Hughes.

and the supermarket where Mary buys her canned stuff—is full of large families, Polish and Irish, Jewish and German, Swedish and Italian, with no signs of friction between either the trooping youngsters or their polyglot parents. "If any of ours get in a row," Mary says, "they daren't come back telling us about it—they have to go back and fix it their own way by themselves." Big oaks and maples and horse chestnuts make Madison Avenue a pleasant street, although the houses are mostly plain gray boxes that would be dreary enough in themselves. Even the huge gray gas tank across the way, which horrified Mary when it was built five years ago, is not the eyesore it might be. In Fred's golf-club days the neighborhood was only a three-mile drive from his work, and then Mary's Royal job was just a step out the back door, so they have been here seven years. It takes her half an hour now to get to her job, where the whistle blows at seven.

Five rooms on the ground floor of this narrow gray house definitely cramp the Berckmans these days. Three little girls in two big beds in one bedroom, a small bed for Junie in his parents' room is the way they manage it. The oil-burning half of the big gas-and-oil kitchen range heats kitchen and bedrooms and the oil heater in the dining room takes the chill off the rest. When the chill refuses to quit the living room of bitter winter evenings, Mary and Fred move their rocking chairs out near the stove and listen to the kitchen radio.

Very quiet people, friendly but not intimate with neighbors, not inclined to be joiners. Fred let the American Legion slide and Mary's only membership in anything is the Ladies of la Salette, a church organization. On Sunday Fred takes a youngster or two to the movies—all of them are just as fond of horse opera[4] as he is. Tuesday Mary has her movies, also with a juvenile escort, usually looking for the kind of good laugh that she gets out of Abbott and Costello. By now constant attendance at the local theater's "dish night" has almost completed a set of pretty blue-and-gold china carefully stowed in the dining-room closet for best. Otherwise they stick domestically to the radio, which Fred turns on right after supper, listening equitably to anything that comes along—news, comedy or symphony, it's all the same to his grave attention.

By the back porch stand receptacles for the tin cans and other salvage that the youngsters are eagerly collecting for Uncle Sam. On the living-room wall, along with family photographs and "Souvenir of Ireland" post cards, now hang two glass plaques that Mary bought from a street peddler—a portrait of General MacArthur and a view of a battle-

---

[4]Western movies.

ship passing the Statue of Liberty, with the legend, "Remember Pearl Harbor."

They aren't likely to forget. Fred had been vaguely figuring the nation would get into the war somehow, sometime, but the way it happened was shattering, and Mary, she says, hadn't given the war a thought:

"But I've been boiling ever since. I'd slept late that Sunday and gone late to Mass and we were working away here at home, baking a lot of stuff, and the radio wasn't on. I had to run out to the store for a loaf of bread and the woman said to me, 'They're killing our boys in Hawaii. Turn on your radio!' I ran home to Fred and he didn't believe it. I switched on the radio and he listened and then he said, 'I guess I don't feel like going to the show today. I feel sick.' We all did. The kids were scared, sort of trembling. You'd have thought the war was right outside. And at the plant next day we could hardly do any work, we were all so mad and disgusted. We felt like throwing things. Ah, I'd like to get at them!"

She is getting at them. With every odd penny dropped in the little bank that is periodically opened to buy more War Stamps. And with every gleaming cylinder that passes through her hands to be incorporated into side arms for fighting men. "When you're in war work," Mary Berckman says proudly, "you're in the Army now!"

6

J. EDGAR HOOVER[1]

"Mothers . . . Our Only Hope"

Woman's Home Companion, January 1944

Every woman should read these pages with the utmost attention. Probably the most important crisis on the home front today is juvenile delinquency, now risen to alarming heights. Here two prominent authorities—J. Edgar Hoover, head of the F.B.I., and James Madison Wood, girls' college pres-

[1]J. Edgar Hoover (1895–1972) was director of the Federal Bureau of Investigation from 1924 to 1972.

J. Edgar Hoover, "Mothers . . . Our Only Hope," *Woman's Home Companion,* January 1944, 20.

ident—explain how mothers can and must put their responsibilities to their children above all else. Recently we published a Companion Poll in which our readers endorsed by an overwhelming majority sex education in high schools as a weapon to combat juvenile delinquency. In early issues we shall publish other forthright articles on this vital problem affecting the America of today—and tomorrow.

—THE EDITORS

At the F.B.I. we try to be realistic about what is known as juvenile delinquency. We even call it by its right name.

A certain boy at fifteen was an honor student and an eager reader of the best books and magazines, healthy and athletic and in every way normal. A year later he is driven first to perversion and then to crime. When we see this same boy at sixteen beginning to serve a life sentence for the brutal murder of his little six-year-old sister—all because of parental incompetence and neglect during a critical period in the boy's development—we know that what we are up against, what the whole country is up against, is not juvenile delinquency but adult delinquency.

We have also learned not to blame the upsurge of youthful crime entirely on the war. That is too easy and too untrue. Adult delinquency on a nationwide scale preceded Pearl Harbor by at least two generations. The war merely aggravated it.

Girls and boys who are now mothers and fathers suffered from adult delinquency of the past. If they allow the disintegrating process to continue until they and their own children are completely isolated one from the other, it is because they have never been taught how to do otherwise. They are themselves second-and-third-generation delinquents, adult in years but not in parenthood.

This progressive fading of parental responsibility has resulted in the so-called "wiseness" of the modern child. He refuses to take his parents as models. He is not only "wise" to their deviations from the straight and narrow, but suspects deviations, even takes them for granted when they do not exist. So he chooses his own models outside the home—often, as we know, with the unhappiest results. This tendency can and must be stopped by the parents themselves. If they do not know how to do it, they must be taught.

At this point the war enters the picture by placing a double burden on mothers. Fathers who might be roused by a campaign of education to become not only examples but preceptors[2] for their children are in many instances away at war or at war work; or they are employed at hours which make normal family contacts impossible. If the drift of normal

---

[2]Teachers, mentors.

youth toward immorality and crime is to be stopped, mothers must do the stopping.

They must not only feel responsibility for setting the right example for their daughters, but they must keep continually before their sons' eyes the good examples of their absent fathers. More than ever they must find ways to teach their daughters that chastity is the best policy, and they must steel themselves to banish false modesty and give their sons the advice and knowledge which will bring them through this trying period clean and well.

They, upon whose shoulders wartime deprivations fall most heavily, must feel a double obligation to bear without resentment the pinch of rationing and the annoyance of temporary interference with private life. They must not give even the appearance of evading or disobeying these necessary governmental regulations lest they encourage their children in evading and disobeying necessary parental regulations. They must set examples in patriotism as well as morality.

But when is the already overburdened mother, especially if she is a war-worker mother, to find time for these double duties?

What practical program can she adopt?

In the first place, unless family finances absolutely demand it, the mother of young children should not be a war-worker mother, when to do so requires the hiring of another woman to come in and take care of her children.

Hard pressed as our manpower authorities have been, they have adhered steadfastly to the principle that patriotism does not consist in one person or group of persons giving up duties which only they can perform to assume different duties which others can perform just as well or perhaps better.

This wise governmental policy applies with special force to mothers. Motherhood has not yet been classed as a nonessential industry! There is small chance that it ever will be. The mother of small children does not need to put on overalls to prove her patriotism. She already has her war job. Her patriotism consists in not letting quite understandable desires to escape for a few months from a household routine or to get a little money of her own tempt her to quit it. *There must be no absenteeism among mothers.*

That last sentence should, I believe, be taken literally. It is the essence of the whole program. The happy home—the one in which there is no delinquency, no matter what adjective you wish to place in front of it— is the home where the child rushes in and calls, "Mother!" and gets a welcoming answer. If, as so often happens in wartime, that child has done a day's work, Mother should be on the job to give that answer.

To back it up, there should be a hot meal ready to serve and a mother fully dressed and ready to receive not only her own children but their friends. They won't be the friends she used to know, whose mothers she knew. Those old boy friends—and girl friends too—are apt to be far away. But the girl whom her son picks up in the factory commissary or the soldier her daughter meets without the aid of introductions at the corner drugstore may be equally decent if they are given a decent place in which to prove it. And that is for a mother to provide.

The mother who does not provide that decent place is definitely falling down on her war job. Whatever rearrangement of her own eating, sleeping and working hours is entailed, she must be ready to give her children and their friends—no matter of how recent vintage the latter may be—hospitality and decency. If she fails to do this she is driving them to places of their own choosing, clandestine places, where there may be hospitality but where decency is unknown.

If her burdens are already too heavy or her strength too frail to permit her becoming a two-shift or three-shift mother, she must find some way of staggering these emergency duties among relatives, neighbors, friends. This applies especially to the comparatively small group of mothers who must take jobs which keep them from their homes at the hours their children most need them, and to the unfortunately larger group whose families live in such cramped quarters that both old and young are driven onto the streets and into the taverns during those hours when the family wage-earners must sleep.

If all else fails the wise mother will utilize one of the many public or private agencies which are now doing such a good job in helping to meet this problem by providing, not homes of course, but respectable gathering places where wartime acquaintanceships unconventionally begun can at least be conventionally continued. Manpower in these agencies is still short, but volunteer service by men and women whose hearts are in the right place has now taken up much of the slack and lent to the work a new spirit and easy warmth which in the past it has sometimes lacked.

In short, the situation is far from hopeless for any mother who really wants to do a good job for her children and will give time and thought to working out a practical program for doing it. And if enough mothers do give that time and thought, the situation is far from hopeless for the nation and its youth.

7

JAMES MADISON WOOD[1]

"Should We Draft Mothers?"

Woman's Home Companion, January 1944

We face a crisis on the home front which soon may require the drafting of all American mothers. I refer not to the manpower shortage in our vital defense industries but to the growing shortage of mothers in hundreds of thousands of American homes.

This absentee-mother problem already has produced the most critical juvenile delinquency situation in our entire history. The situation becomes more acute every day as more and more American mothers swap their aprons for overalls and trade their homes for auto-trailers.

Many of these mothers naturally are attracted by the pay checks promised them by the defense industries. Others are sincerely trying to do their patriotic duty, as they have been misled into seeing it. They are told at every turn that only in the factory can they help win the war.

You hear so many mothers say, "I feel guilty because I'm not working in a war plant." These mothers have been led to believe that the rearing of children has become a nonessential activity in this country. Apparently they do not realize that their primary responsibility is to their children and the America of tomorrow.

The job of every American mother, in war and peace, is to care for her children. There is no more important job than that in this country today. The future of our nation depends every bit as much upon the morality and health of our next generation as it does upon the winning of battles in Europe and Asia. Schools, nurseries and playgrounds can help train our youth but they are no substitute for the home.

The American home, whose only guardian in wartime may be the mother, is the very foundation of our way of life. Military victory will be

---

[1]James Madison Wood (1875–1958) was president of Stephens College in Columbia, Missouri, then a two-year college for women, from 1912 to 1947.

James Madison Wood, "Should We Draft Mothers?" *Woman's Home Companion,* January 1944, 21.

a hollow mockery if we produce a generation of demoralized and delinquent Americans while winning this "war to preserve our way of life." Your daily newspaper and every local police record will show you how serious this threat has become. We have thousands of underfed neglected children tied to clotheslines, locked in cellars or left to run wild, while Mother wields a blowtorch. We have courtesans, schoolgirl age, diseased in mind as well as body. We have more illegitimate babies.

Our juvenile delinquents are learning every form of vice and crime — drunkenness, dope addiction, murder, kidnaping, rape and robbery. While we are fighting the international gangsters, we are breeding in our own midst countless young gangsters who might have been decent useful citizens if their mothers had not put activities outside the home ahead of their duty to their children and society.

Essentially I am opposed to drafting American women for war work. I believe that the women of a democracy should be able to recognize their responsibilities and act accordingly. But if they fail, mothers should be drafted — and ordered to accept their responsibilities for the protection of society. They should be assigned to duty in their own homes. I would do this in a very special way. I would draft them first. I would tell them they were being called first because their job in their home is as important as any other job in the nation. I would see that they were recognized as being fully as vital to the war effort as our women in uniform or overalls.

If the draft imposes financial hardship on some mothers, why not pay them for staying at home and rearing the America of tomorrow? As a nation we could make no better investment. If we are willing to pay for reform schools, penitentiaries and homes for illegitimate children, we should be even more willing to pay for the support of the American home. Actually the plan would pay for itself by saving the millions which juvenile delinquency now costs us.

But is anything being done to control this juvenile delinquency problem? Federal and local police are enforcing the law but that is no cure.

Mothers are still being lured from their homes with pay checks, are still being told that their patriotic duty lies in the factory. But no provision has been made to compensate for their absence.

Instead of having more playgrounds, more schools, more nurseries, we have fewer. The federal government has made an attempt to increase recreational facilities for factory families, but local communities have been slow to cooperate. Many playgrounds and schools have been closed in the name of economy.

Our uniformed services for women accept no mothers with children under fourteen, but here again we see only a partial understanding of the juvenile delinquency problem. Children between fourteen and eighteen are the very ones who need parental attention the most. More of them are appearing in police line-ups today than any other juvenile group.

When we convert a wilderness into a teeming defense plant area, usually the last things to go under construction are the workers' homes, playgrounds, nurseries and schools. The result is hardly surprising. Juvenile delinquency is far higher in these war factory areas than elsewhere.

We here at home, who claim to be backing up the boys at the front, actually have been undermining the very thing those boys fight for. They want to see the American home maintained and protected. They want to see the next generation of Americans strong in character and high in morale. That is the only hope of avoiding another war.

This is the basic responsibility of the American mother: she is the only one who can turn the tide of juvenile delinquency. She is the only one who can guard the home. Her patriotic duty is not on the factory front. It is on the *home* front!

8

ALFRED TOOMBS

"War Babies"

Woman's Home Companion, April 1944

Must they be neglected and badly fed? Of course the answer is no, and again no! But what are the facts?

America may be on the way toward creating another lost generation.[1] We run the risk by paying too little attention to the welfare of our war babies—the helpless creatures now being born into a topsy-turvy world.

[1]The term "lost generation," coined by author Gertrude Stein, refers to the generation of Americans who came of age between the two world wars. Their rootlessness and hopelessness, exemplified in the lives of expatriate American writers living in Europe, were chronicled by those writers, notably F. Scott Fitzgerald, Ernest Hemingway, and Sherwood Anderson.

Alfred Toombs, "War Babies," *Woman's Home Companion*, April 1944, 32.

That is the opinion of many government officials, social workers, psychiatrists and public health experts with whom I have talked in recent weeks. The authorities are particularly alarmed by what is happening to babies whose mothers are working or living in war production centers and to babies whose fathers have been called off to war. And cause for alarm has not been materially lessened by the recent increase in the allotment to servicemen's dependents.

Social workers have reported that cases involving the neglect of infants are increasing twice as fast as cases involving juvenile delinquency. Nevertheless while we are taking steps to meet the juvenile delinquency problem, the nation as a whole has failed to face the problem of neglect.

I have seen something of the conditions which are affecting the lives of our war babies — and I understand the alarm felt by officials.

On tiptoe, I peered through the window of a trailer at the edge of a war-boom city recently. A baby lay in a crib within, moving his head listlessly from side to side. A social worker with me said:

"The neighbors say that the baby is left alone like this every afternoon when the mother goes to work. Her husband is in the service and she works on a late shift in town. She has no one to watch the child, so she just leaves it."

In the same city I visited shabby rooming houses where I saw underfed babies sharing their cribs with vermin. I went to war housing projects where infants less than two years old were left under the loose supervision of neighbors — or under no supervision at all — while their mothers worked. One baby I saw had been left alone in the yard the day before, her only shelter a play pen. A storm had come up and only the thoughtfulness of a neighbor had saved the baby from prolonged exposure. But the day I saw the baby it was out in the yard again — alone.

No question is of more vital interest to the women of America than the welfare of our babies. I have read with interest the article written for the Woman's Home Companion by Alfred Toombs. I recommend that when you finish reading it, you ask whether your community has met the problem he discusses. If it has not, why don't you see what you can do to help?

KATHARINE F. LENROOT
Chief, Children's Bureau
U.S. Department of Labor

The community I visited obviously had failed to take any determined steps to protect the welfare of the babies I saw. And although some communities have worked out a solution for the problem—proving that it can be done—most have not faced up to the facts.

We are in the midst of the greatest baby boom in our history. At the same time, normal patterns of family life are being destroyed. Men are being called into military service or drawn into crowded war factory areas where normal existence is impossible. Women are going to work at an astounding rate—eight million are believed to have left their homes since 1940 to take jobs. A substantial percentage of these are mothers. And the fact—which must be faced—is that America's babies often become victims of this chain of circumstances.

That the fact is not faced is seen in reports from various parts of the country. Appalling neglect and mistreatment of infants seem universal. The most horrible reports tell of babies dying in fires in unsupervised nursing homes or being beaten and otherwise abused by their daytime "guardians." In at least three different places, groups of children were found herded into a chicken-wire enclosure in the basement of a house while their mothers were doing war work.

In a southern city, investigators found an infant left in a rooming house to be cared for by anyone who took pity on it while the mother was working. In a near-by city they found a baby which had been left in its crib for twenty-four hours without once having been picked up.

Children born to the wives of men in the armed services have been among the chief sufferers. It is estimated that more than half a million men in service have small children. Until last fall, these children were given only a pittance by the nation. Congress, hearing reports of distress, raised the amount of the allotment granted servicemen's dependents. A wife is allotted fifty dollars a month for herself, thirty dollars for one child and twenty dollars more for each additional child. But the Bureau of Labor Statistics estimated that it would take eighty-four dollars a month to maintain a wife and one child on a minimum budget. The steady rise in the cost of necessities has made it increasingly difficult for the serviceman's wife and child to live on eighteen dollars a week.

The desperation of some of these service mothers is reflected in reports which reach government officials. In an eastern seaport, police found a child of less than one year abandoned. Eventually the mother was located. She was the wife of a navy enlisted man. She told investigators she had been unable to support the child on her husband's pay and had no family to which she could turn. Since she could find no one to care for the child, she hadn't been able to work. In a frenzy of fear, she had abandoned the baby.

The Children's Bureau receives many pleas for help from young mothers. The Bureau supervises the most successful program that has been undertaken to help service mothers. This is the emergency maternity and infant care program, under which the federal government has made funds available to the states to provide medical care for infants and pregnant wives of servicemen. Wives of enlisted men in the four lowest pay grades receive free prenatal care, delivery and postnatal attention and pediatric care for their infants up to one year of age. A hundred thousand young mothers have benefited from this program—an example of a method by which other problems of war babies can be solved.

In cities where there has been a sudden influx of service personnel, conditions affecting infants are usually aggravated. Near a North Carolina town—the population of which has increased from seven hundred to twelve thousand—women with small children were living in garages and chicken coops. Public health nurses near Fort Leonard Wood, Missouri, reported infants suffering from malnutrition and illnesses caused by poor living conditions. Admittedly, women should stay out of such places, but there they are, by the thousands, and with them their babies.

The service mother who finds it impossible to live on the government allotment also finds it hard to go to work because facilities for caring for her infant are inadequate. In Washington, D. C., a young navy wife found she had to work and appealed for help in caring for her baby. But the public agencies could not help her—they had had more than seven hundred applications for infant care during the previous year and had been able to fill only one third. A soldier's wife, encountering the same problem, went out herself and found a woman to board the baby. The woman left town, leaving the infant with her sister, who lived in a basement with seven children of her own. The mother found the child underweight and ill—and gave up her job.

Foster Homes Recommended

In spite of the fact that the War Manpower Commission has tried to discourage mothers of young children from working, thousands have taken jobs. It is estimated that six per cent of the working women have infants and in California the number runs as high as ten per cent. The federal government has provided funds to help establish nursery schools. But children under two years of age need individual care that is difficult for nursery schools to give. The Children's Bureau recommends that where mothers must work their infants be placed in foster homes during working hours. In such homes, carefully supervised, a foster mother can care for two or three infants. But federal funds are not available to finance a program of foster home care.

Last summer there were six hundred and sixty-two war areas in the country where labor and living conditions indicated a need for child care programs. Only sixty-six were providing any sort of infant care.

The lack of day care facilities has not prevented all young mothers from going to work—but rather has resulted in the neglect or mistreatment of some babies. Testifying before a Senate committee about conditions in the San Fernando Valley of California, Mrs. Louise Moss, of the Office of Civilian Defense, said:

"I have seen children locked in cars in parking lots in my valley and I have seen children chained to trailers in San Diego. Those stories are not fiction. I have seen that myself. But no one has raised a voice long and loud enough to stop these conditions from going on month after month at a fast-increasing pace."

An infant needs the love and care of its mother or at least of some one other person who can be a mother substitute. But many working mothers are forced to make haphazard arrangements with neighbors and relatives which are almost certain to be harmful to the child. An infant passed from hand to hand for care day after day may develop a sense of insecurity which could result in nervous disorders, child psychiatrists told me. One psychiatrist told of a mother she knew who had made this arrangement for her baby: two days a week a maid cared for it; three days a week, neighbors watched it and on Saturday an older child looked after it.

In Norfolk, Virginia, I saw the sort of conditions which prevail in many crowded war centers. Hundreds of mothers and infants had poured into the area. Some were wives of servicemen, others brought in by the humming war industries. They lived where they could—in cheap rooming houses, trailers or slums. They went to work, making catch-as-catch-can arrangements for the care of their babies. The result was a serious social problem.

Before the war child welfare authorities in Norfolk handled annually about a hundred cases involving neglect or dependency of infants. In 1942 there were more than three hundred. These cases involved outright abandonment, babies beaten black and blue by those supposed to care for them, babies suffering from skin diseases acquired in filthy rooms. Virtually all the children involved had to be hospitalized. And for every case which came to the attention of the authorities, it was suspected that there were a dozen which didn't. Norfolk was making an attempt to establish a foster day care program for infants, but the undertaking was progressing slowly. Welfare organizations in the city were woefully understaffed. Norfolk needed more workers, professional and volunteer. In many war areas, conditions are bad enough to affect even the well-being of infants whose mothers do stay home to take care of them. Housing conditions and the doctor shortage present health menaces. Recent figures on infant mortality rates are not available, but officials say it may be a few years before present conditions would be reflected in statistics. A child suffering from poor nutrition or disease may not succumb immediately. But its chances to reach maturity are lessened.

The authorities to whom I talked agreed that the conditions which menace the health and welfare of so many babies today can be corrected—as soon as there is public determination to act. They pointed out that some communities have already taken steps to meet the problem.

A first step toward providing infant care has been the establishment in some cities of advisory and counseling services. Many have been organized by local defense councils and receive the support of other community organizations.

In Rochester, New York, officials became aware of the plight of small children and began in earnest to seek for foster homes. A questionnaire went out to all women, asking whether they wanted to work or would be willing to look after the child of some working mother. More than seven thousand women offered to care for others' children. Nearly five hundred were interviewed and three hundred were accepted to care for small children.

Adequate Funds Needed

Limited federal funds are available to the states to aid in establishment of day care centers for children over two years old. Measures to provide money to promote infant care programs and other services necessary in an adequate child care program have been pending in Congress for some time. If local communities could be provided with adequate funds, they could solve the problem. The money could be used to establish information centers for mothers who seek day care for infants. It could be used to hire staffs of social workers to locate and supervise foster day homes. Some states and cities have made progress by financing these programs themselves, but experience indicates that federal grants are needed to insure complete success.

Many believe that further consideration should be given to the problem of allotments for wives and children of servicemen. It has been urged that, beyond the flat allotment now granted, there should be additional funds available for hardship cases. A young service mother can find shelter for herself and her infant with relatives and live without too much privation on eighty dollars a month. But what if she has no relatives? Shouldn't there be some provision for granting her more than eighty dollars a month?

And no one will question the necessity for cleaning up living conditions around war centers. Adequate housing, medical facilities and food must be made available in these places. These are problems which can be solved through community action.

These things can be done and must be done—lest in our haste to destroy our enemies we injure ourselves.

9

"When Your Soldier Comes Home"

Ladies' Home Journal, October 1945

At first, he may find it more difficult to live with you than without you.

As a former front-line infantryman, returned because of wounds incurred during the Battle of the Bulge, I've read a spate of articles dealing with the problems of readjustment for the returned veteran. I compare them with my own experiences and those of my friends. It makes me wonder how people who haven't gone through what we have can write so glibly about it. These articles carry the easy assurances from experts whose eyes are closed to the variety of human nature and the complexity of modern war. They write as if all men return with similar reactions and face similar problems.

Today the word "soldier" is so inclusive that it covers both sexes, and a soldier's job is divided into thousands of specializations other than fighting. For a very few the war meant forward foxholes, house-to-house fighting or combat missions. For most men it meant handling supplies, checking reports, driving vehicles, drilling recruits, building roads, repairing equipment, working in offices and a host of other necessary functions.

The odds are that your soldier won't come back from the war with horrible memories. He is likely to shrug off inquiries about combat and feel a little guilty and frustrated, as if having been cheated out of a great experience. As with most soldiers, war for him has meant drudgery, not heroics.

If it is not the strain of combat, then what makes readjustment to normal life so difficult? The answer is: Having been a soldier, he finds it hard to be a civilian. Having missed you greatly, he has been forced to find ways of getting along without you. Having loved his fellow soldiers and hated the Army, he is lost when turned free as a civilian. Having looked forward so long and earnestly to civilian life, he finds it dull stuff compared with his dreams.

After the first few whirlwind hours, the actual problems of readjustment begin. It is a little like the letdown that follows the honeymoon, when jobs and finances compete with hugs and kisses. Your man has grown apart from former friends who have not been to war and, among a crowd of civilians, he searches for G.I.'s. It is as though, underneath the new civilian suit, he wears an invisible khaki uniform.

The transition is difficult for him because he feels lonely. For the first time in years, he is not surrounded by hundreds of other men who are very much like himself. He met the difficulties of basic training with comrades. Together with them, he faced the war and sweated out his discharge. Now that he is out, he is alone. And in that terrible loneliness, the past is touched with a softer light. He senses a new, mocking truth in the ironical barracks advice: "Quit your kicking, soldier. When you get out and become a civilian again, you'll never find it as easy as in the Army. You found a home in the Army."

This prophecy, which satirizes every G.I.'s desire for discharge, now assumes a partial truth. He discovers, upon being separated from the service, that he had found a home, that he had belonged to this special world whether he liked the Army or not. In the months that follow he learns, as all of us do, that the way of the returned soldier is hard and lonely.

You must understand that his conception of civilian life has all the glittering novelty of holidays and none of the drab routine of workdays. The postwar world he has imagined is essentially a retreat from the past years of regimentation when, by contrast, civilian life appeared so desirable. To a soldier, it is a special sort of existence, free from regulations and not responsible to an alien will. And because these privileges are denied him, the soldier feels distant and somewhat hostile toward others. His world divides into two groups: "the guy who's in and the guy who's not in." The latter lives off the soldier and prospers through his adversity. Civilians are selfish, grasping and complacent.

Now that he has changed sides, the ingrained values apply to himself. His conscience begins to work as he remembers that others are soldiers while he is free. No matter how much action he may have seen, nor how much he may have sacrificed, he feels guilty. A soldier in time of war always feels that he is doing less than others. If he is in this country, he feels guilty because he is not overseas. Back in rear echelon, he feels guilty because he is not in the front lines. In the front lines, he feels guilty if the sector is quiet. After a fierce engagement, he feels guilty because he has survived, while his friends were killed, or because he was wounded and evacuated while they remained. When he comes home for good, discharged on points or wounds, inside him that guilty feeling is reinforced by a sense of loneliness.

He wants a special sort of comradeship which cannot be found outside the society of fighting men. No matter how much you mean to him, you cannot provide that. He must repress this loneliness and struggle along for a while without this comradeship until the desire for it gradually dies.

"No matter how difficult it will be at first, we'll make out all right," you wives and wives-to-be exclaim. "We love each other, and we'll work it through."

And that's likely to be the truest prophecy. Your love can be the anchor that steadies him during the stormy period of readjustment. But you must meet it with more than blind love. You must have patience and understanding. Even the intimate character of his love for you has been affected subtly and deeply by the long separation. Your man who found it so hard to live apart from you may find it almost as difficult to live with you again. The soldier, adoring and worshipful when away, may be a perplexed and unwilling Romeo when home again. Somewhere along the line he has become unduly sensitive. You'll notice his tendency to fly off the handle and say things to you that he doesn't mean.

Part of this friction you can rightly attribute to the fact that he is feeling the strain of readjustment and is "taking it out on you." A measure of that you can understand and forgive, but there will be times when he is with you, but his thoughts are a million miles away; he seems to withdraw within himself. You almost suspect that he doesn't want you any more. It's then that you need to have a deeper comprehension of what lies behind this unwelcome turn in your relationship.

Your soldier has been away from you and from the American community for a comparatively long time. His position in an all-male world and his fleeting excursions into the civilian life of other countries have given him little preparation for a return to close association with the distaff side. You, as an American woman, have a higher status and more importance than the women of any other country; and to your G.I., that is not a dry sociological fact. He has spent time in another world where men rule supreme and where women are marginal creatures at best. It will be difficult at first for him to yield to so many feminine influences.

He is not accustomed to more than intermittent contact with female society and so, when you suddenly burst upon him, however welcome you will be initially, he will shortly reach a saturation point.

As a result of the long period of absence, he is not prepared to resume the close partnership with you as an equal and to accept you without reservations. He will tend to resent inroads on what little privacy he has managed to safeguard against years of Army effort to make him an interchangeable unit. From force of habit, he will guard his affections and private thoughts closely. His long-checked emotions are not released easily again just by removing the uniform. He has been independent of you for so long that sharing his life with you comes hard at first.

While he has been separated from you, he has thought about you often and missed you deeply. Out of that longing and desire has grown an image which is not you so much as an idealized version of you. Since he could not have you in his everyday world, he has enshrined you on an impossibly high pedestal in his dream world.

You would be flattered but a little scared to know the complete extent of his worship. Your picture is in his wallet and taped inside his helmet. As a doughboy,[1] he gives his rifle your name and fondles it. If a tanker, he paints your name on his turret. Wherever he goes, you go beside him. Your picture comes out frequently, not for his own inspection (he won't forget you), but for the benefit and enlightenment of his buddies. It gets to be a case of, "I'll admire yours if you'll admire mine." But privately he adds, "Mine's really much nicer."

There you have been for these many months — a perfect being in his imagination. But the separation has been long and, even if you hold an undisputed claim to all his affections, what about less ethereal substitutes?

Well, you have them. That is the cold truth. It is the rare soldier who can honestly say that he has never, even in any degree, strayed from the straight and narrow path into the regions you would place "Off Limits" to him. The G.I. slang for it is "liberating a girl," and it means just what you suspect. Remember those pictures of cheering girls who bestowed their affections on the victorious Yanks in Paris and Rome? There were a lot of towns and villages in occupied Europe where the Yanks were just as warmly welcomed by women who hadn't seen their own men for years. To them your soldier looked pretty good. Well, why not? He looks pretty good to you, doesn't he?

Your soldier met a very ardent public which embraced him literally as well as figuratively. He brought not only freedom but also chocolate, cig-

[1]Colloquial term for an American soldier, first used during World War I.

arettes, food and American gallantry. And they showed their gratitude in diverse ways.

In the soldier's unwritten code of morals, these lapses are perfectly excusable. The hedonistic philosophy, "Eat, drink and be merry, for tomorrow we die," is perfectly suited to military life. Were it not for the satisfactions offered by these occasional flings, his life would be more difficult. Through these forbidden adventures, he preserves the remnant of a private life and gains an outlet for repressed urges and new patience to go out again. If his conscience troubles him he thinks, "What the hell, I may as well get some fun out of life. I may be killed soon. And it won't hurt Her anyhow."

Even if there is little possibility of being killed, he acts as though there were. As Kipling said of other soldiers in other wars, "Single men in barricks [sic] don't grow into plaster saints." Nor do married men either. In fact, G.I. philosophers believe, "Some married men are the biggest wolves. They know how to howl the loudest."

So, missing you and being very human, your man kicks over the traces occasionally. No one else is there to make a fuss over him, so when one of the local females indicates that she thinks he is charming, the fall is imminent. But with characteristic American values, he says to himself after the conquest has been completed, "Thank God my girl back home isn't like that."

If you're worried about losing him to the charms of liberated belles, relax and rest easy. You don't need to be jealous or threaten him with reprisals. Take it as a compliment that most of these short-lived adventures were unsatisfactory substitutes for the deeper, more meaningful life he has known with you. They don't endanger you, nor should you make the mistake of inventing imaginary flirtations of your own to convince him that two can play at the same game. After all, he had the excuse of a soldier's life while you have none. Besides, you don't want to drive him away by putting yourself in the same class with his temporary amours. They are only lesser rivals compared with that glamorous, idealized version of you that he has been carrying around in his imagination. Your chief rival is a goddess-self—the image he has created in his memory of you. Your biggest problem is to make the flesh-and-blood woman as interesting to him as that image, while yielding nothing by way of allure to those other women he has known.

As a mere man, I don't have to go any farther and tell you how to win him back or hold him. You will know how to go about that business yourself. After all, you landed him once, and he'll probably be more susceptible than ever.

The chief point by way of warning is: once you've got him, don't try to hog-tie him in place. For a long while memories of past events will be with him, and he will leave you in his thoughts to renew those former times.

Figure 2. Cocomalt advertisement from *Ladies' Home Journal*, May 1940, 111.

Those war years are his; and for the present, when he thinks of them and of his comrades, he will be distant from you. During those reveries, let him go. He will come back.

The countless little things that are symbolic of his Army past—and you will learn them by experience—hold a sacred place in his sentiments. The buddies of those years, the souvenirs he has collected, the personal Army habits he has maintained, the little incidents he never tires of retelling—all these he will value. They are the links that bind him to a past that will become glamorous and exciting after the more proud elements fade.

Don't belittle him if his adventures were not very martial or urge him, if he did know combat, to recount his battle experiences before an audience of your friends. Remember what they mean to him, and remember that no matter how proud he may be of his participation in the war or of his decorations, the soldier's conscience within him says, "You've come back while others, braver and more deserving, have died."

It is this voice that makes him sensitive and reverent. It is this that links him with the gallant men, living and dead, who fought the war. Do not begrudge him these minutes when you feel him by your side and yet apart from you, for then he is back in the war with his comrades beside him and it is good for his soul.

When he tries to tell you these things, take it as a high compliment. If he has been in combat and wants to talk about it, as most soldiers do, listen to him carefully and attentively, for he is sharing a great experience with you. The more he tells you, the greater his trust in you. He is enlarging his sacred circle of fighting comrades to give you admission. He is telling you about these unforgettable days when, like Oliver Wendell Holmes, Jr., his "heart was touched with fire."

Remember these things as he travels the slow road back from soldier to civilian. As you love him, be patient with him during the trying times. Remember that, just as it took work and patience to make your marriage, so it takes more of both to resume it now.

Remember—you were once a goddess.

2

Women and the Workplace

The mass-circulation "service" magazines such as *Good Housekeeping, Ladies' Home Journal, McCall's,* and *Woman's Home Companion* always considered their readers to be women who worked—worked, that is, as homemakers and sometimes as volunteers in their communities. While these magazines frequently published articles about women who indisputably had careers outside the home—media celebrities, health professionals, journalists, and even politicians—these were clearly presented as exceptional women, not necessarily role models for the average reader. Both during the war years and afterward, the magazines mirrored the culture's deep ambivalence about women's paid employment. Titles of two articles in the January 1944 issue of *Ladies' Home Journal* announce the two sides of a continuous argument: "You Can't Have a Career and Be a Good Wife" and "Working Wives Make the Best Wives." It is not surprising that in magazines aimed primarily at homemakers, the majority of articles on the subject warn of the dangers of trying to inhabit both worlds: fatigue, neglected children, husbands with threatened egos. A *Journal* poll in 1940 indicated that its readers agreed: 74 percent of those responding believed that housewives led "more pleasant lives" than did women who worked outside the home. Yet by 1948, in part because of the increase in women's paid employment during World War II, a Census Bureau study found that seventeen million women were in the paid labor force, half of them married.

Other women's magazines, especially those intended for younger readers, were less focused on woman's role as homemaker and thus were more supportive of women seeking paid employment—although usually in areas traditionally regarded as appropriate for women: as secretaries, models, teachers, and airline flight attendants (then called stewardesses). *Seventeen* devoted much of its July 1954 issue to young women involved in the theater; *Mademoiselle* regularly offered quite specific advice on obtaining entry-level positions, addressed to the college student or recent graduate. A 1953 *Mademoiselle* article even advised women on how to pay gracefully for a business lunch with a male peer. Yet articles in these same magazines concerning fashions, dating, cooking, and home decoration suggest that a woman's true vocation was marriage and, further, that she would trade her typewriter for a rolling pin when she wed — indeed, flight attendants were required to be single.

Social class differences show up more clearly outside the orbit of the magazines for homemakers. Readers of *Harper's Bazaar* and *Mademoiselle* were presented with models of volunteer work: in the former magazine, planning charity balls, and in the latter, becoming members of the Junior League. At the other end of the spectrum, the December 1953 *Coronet* (the magazine for "the American family"), in a rare magazine nod to the existence of African Americans, profiled a southern black woman whose diligence had transformed her from sharecropper to prosperous landowner.

To whatever extent a woman occupied herself outside her home, the magazines clearly assumed that she was primarily responsible for what went on inside it as well: cooking, cleaning, child care, and the purchase and disposition of household goods. In the world presented in the magazines, men might mow the lawn, build shelves in the garage, or grill hamburgers on the patio; they did not run vacuum cleaners, diaper babies, or bake cakes. The female college professor who asserted in a 1949 issue of *Ladies' Home Journal* that working couples should share equally "all family responsibilities" was an unusual voice. Far more typical was the article in the *Journal*'s "How America Lives" series dramatically titled "I Gave Up My Career to Save My Marriage"; the article describes a woman abandoning her aspiration to sing with the Metropolitan Opera to stay home with her four children.

In the climate of the mid-1940s, the Equal Rights Amendment to the Constitution, which had first been proposed in 1923, seemed even more of a threat to the status quo than it would in the 1970s; equality of opportunity implied the loss of certain protections for women in the workplace — if they had to be there at all. By the mid-1950s, however, largely because of a consumer culture that encouraged the two-paycheck family, more than a third of women over thirty-five had jobs outside the home, prompting *Woman's Home Companion* to announce, "The Married Woman Goes Back to Work." Perhaps fittingly, the woman's "home" companion ceased publication the following year.

10

MRS. FRANKLIN D. ROOSEVELT

"Women in Politics"

Good Housekeeping, April 1940[1]

Can a woman be president? Will there be a women's crusade against war? Should women have their own political party? These and other vital questions are discussed in this third and final article of the series.

Where are we going as women? Do we know where we are going? Do we know where we want to go?

I have a suggestion to make that will probably seem to you entirely paradoxical. Yet at the present juncture of civilization, it seems to me the only way for women to grow.

Women must become more conscious of themselves as women and of their ability to function as a group. At the same time they must try to wipe from men's consciousness the need to consider them as a group or as women in their everyday activities, especially as workers in industry or the professions.

Let us consider first what women can do united in a cause.

It is perfectly obvious that women are not all alike. They do not think alike, nor do they feel alike on many subjects. Therefore, you can no more unite all women on a great variety of subjects than you can unite all men.

If I am right that, as I stated in a former article, women have caused a basic change in the attitude of government toward human beings, then there are certain fundamental things that mean more to the great majority to women than to the great majority of men. These things are undoubt-

---

[1]This is the last of a three-part series of articles with the title "Women in Politics" written by Eleanor Roosevelt (1884–1962), wife of President Franklin Delano Roosevelt. The Roosevelts were married from 1905 until the president's death in 1945, during his third term. Eleanor Roosevelt was an advocate for social justice whose opinions on subjects ranging from etiquette to politics were sought through her question-and-answer column "If You Ask Me," which appeared first in the *Ladies' Home Journal* and then in *McCall's* during the 1940s and 1950s. In the 1930s, she wrote a monthly column titled "Mrs. Roosevelt's Page" for *Woman's Home Companion.* Eleanor Roosevelt published a number of books, including *It's Up to the Women* (New York: Frederick A. Stokes, 1933), and *The Autobiography of Eleanor Roosevelt* (New York: Harper, 1961). A three-volume collection of her newspaper columns titled *May Day* was published by Pharos Books of New York, 1989–91.

Eleanor Roosevelt, "Women in Politics," *Good Housekeeping,* April 1940, 45.

edly tied up with women's biological functions. The women bear the children, and love them even before they come into the world. Some of you will say that the maternal instinct is not universal in women, and that now and then you will find a man whose paternal instinct is very strong—even stronger than his wife's maternal instinct. These are the exceptions which prove the rule, however. The pride most men feel in the little new bundle of humanity must grow gradually into love and devotion. I will not deny that this love develops fast with everything a man does for the new small and helpless human being which belongs to him; but a man can nearly always be more objective about his children than a woman can be.

This ability to be objective about children is one thing women have to fight to acquire; never, no matter what a child may do or how old he may be, is a woman quite divorced from the baby who once lay so helpless in her arms. This is the first fundamental truth for us to recognize, and we find it in greater or less degree in women who have never had a child. From it springs that concern about the home, the shelter for the children. And here is the great point of unity for the majority of women.

It is easy to make women realize that a force which threatens any home may threaten theirs. For that reason, I think that, as women realize what their political power might mean if they were united, they may decide now and then to unite on something which to them seems fundamental. It is quite possible, in the present state of world turmoil, that we may find women rising up to save civilization if they realize how great the menace is. I grant you that things will have to be pretty bad before they will do it, for most women are accustomed to managing men only in the minor details of life and to accepting the traditional yoke where the big things are concerned.

I have heard people say that the United States is a matriarchy—that the women rule. That is true only in nonessentials. Yes, the husbands spoil their wives; they let them travel and spend more money than foreign women do, but that is because money has come to us more easily in the past and therefore we have spent it more easily. The French woman who is her husband's business partner has more real hold on him than the American woman who travels abroad alone has on her husband. She buys all the clothes she wants without knowing whether her husband will be able to pay the bills, because she is completely shut out of the part of his life that holds most of his ambition and consumes the greater part of his time.

This country is no matriarchy, nor are we in any danger of being governed by women. I repeat here what I have so often said in answer to the question: "Can a woman be President of the United States?" At present the answer is emphatically "No." It will be a long time before a woman will have any chance of nomination or election. As things stand today, even if an emotional wave swept a woman into this office, her election would be valueless, as she could never hold her following long enough to put over her program. It is hard enough for a man to do that, with all the traditional schooling men have had; for a woman, it would be impossible because of the age-old prejudice. In government, in business, and in the professions there may be a day when women will be looked upon as persons. We are, however, far from that day as yet.

But, as I said before, if we women ever feel that something serious is threatening our homes and our children's lives, then we may awaken to the political and economic power that is ours. Not to work to elect a woman, but to work for a cause.

There may be a women's crusade against war, which will spread to other countries. I have a feeling that the women of the United States may lead this crusade, because the events of the last few months have left us the one great nation at peace in the world. Some of our South American neighbors have as much potential greatness as we have, but are not yet so far developed. We women may find ourselves in the forefront of a very great struggle some day. I think it will take the form of a determination to put an end to war for all time.

It is obvious that American women cannot do this alone; but throughout the world this might prove a unifying interest for women. When they get to the point of feeling that men's domination is ruining their homes, then they will use whatever weapons lie at hand.

I think we in this country should be prepared for something of this kind. That is why I said that we must become more conscious of ourselves as women and of the force we might wield if we were ever to have a women's cause. We must be careful, however, not to try to wield this force for unimportant things. If we do, it will split up, for we are as individualistic as men in everyday affairs.

The consideration of future possibilities for peace seem to me of paramount importance; but other things of worth enter into our present considerations. Great changes in our civilization have to be considered, and the women are going to weigh the effect of these changes on the home. I believe women can be educated to think about all homes and not so much about their own individual homes. If certain changes have to be made in industry, in our economic life, and in our relationship to one another, the women will probably be more ready to make them if they can see that the changes have a bearing on home life as a whole. That is the only thing that will ever make women come together as a political force.

Women should be able to weigh from this point of view all questions that arise in their local communities. They should vote with that in mind. But when it comes to standing for office or accepting administrative positions, they should realize that their particular interests are not the only ones that will come up, and that, while they may keep their personal interests, they must prove that as persons they can qualify in understanding and in evaluating the interests of the men, too.

Now let us consider women in the other phases of activity where they wish to be persons and wipe out the sex consideration.

Opposition to women who work is usually based on the theory that men have to support families. This, of course, is only saying something that sounds well, for we know that almost all working women are supporting someone besides themselves. And women themselves are partly to blame for the fact that equal pay for equal work has not become an actuality. They have accepted lower pay very often and taken advantage of it occasionally, too, as an excuse for not doing their share of their particular job.

If women want equal consideration, they must prepare themselves to adjust to other people and make no appeals on the ground of sex. Whether women take part in the business world or in the political world, this is equally true.

A woman who cannot engage in an occupation and hold it because of her own ability had much better get out of that particular occupation and do something else, where her ability will count. Otherwise, she is hanging on by her eyebrows, trying to exploit one person after another, and in the end she is going to be unsuccessful and drag down with her other women who are trying to do honest work.

In the business and professional world women have made great advances. In many fields there is opportunity for them to work with men on an equal footing. To be sure, sometimes prejudice on the score of sex will be unfair and a woman will have to prove her ability and do better work than a man to gain the same recognition. If you will look at the picture of Mrs. Bloomer,[2] made a hundred years ago, and think of the women today in factories, offices, executive positions, and professions,

[2]Mrs. Amelia Bloomer (1818–1894) introduced trousers ("bloomers") for women—originally loose, flowing pants bound tightly at the ankles and worn under a skirt.

that picture alone will symbolize for you the distance women have traveled in less than a century.

In the political field they haven't gone so far. This field has long been exclusively the prerogative of men; but women are on the march. I do not think that it would be possible or desirable to form them into a separate women's political unit. Too many questions arise in government which are not fundamentals that stir women as women. Women will belong to political parties; they will work in them and leave them in much the way that men have done. It will take some great cause that touches their particular interests to unite them as women politically, and they will not remain united once their cause is either won or lost.

I do not look, therefore, for a sudden awakening on their part to a desire for greater participation in the government of the nation, unless circumstances arise that arouse all the citizens of our democracy to a feeling of their individual responsibility for the preservation of this form of government. Otherwise, I think it will be a gradual growth, an evolution.

There is a tendency for women not to support other women when they are either elected or appointed to office. There is no reason, of course, why we should expect any woman to have the support of all women just because of her sex; but neither should women be prejudiced against women as such. We must learn to judge other women's work just as we would judge men's work, to evaluate it and to be sure that we understand and know the facts before we pass judgment.

Considering women as persons must begin with women themselves. They must guard against the temptation to be jealous. That little disparaging phrase one sometimes hears, which suggests that a woman has failed because she is a woman! A woman may fail; but women must begin to impress it upon everyone that a woman's failure to do a job cannot be attributed to her sex, but is due to certain incapacities that might as easily be found in a man.

A good example of the way women tend to follow and play up to a man's opinion came out in a conversation I once overheard. In a hotel corridor a man was loudly proclaiming: "No woman should be Secretary of Labor; she isn't strong enough. These labor guys need someone to knock their heads together." The little woman next to him murmured gentle acquiescence and added, "Miss Perkins[3] has never seemed to me to be a very

---

[3]Frances Perkins (1882–1965) was secretary of labor from 1933 to 1945, the first woman appointed to a cabinet post.

womanly woman." Of course, she did not know Miss Perkins, and, of course, she was only saying what she thought would please her man; but it was a perfect example of our inability as women to think objectively of other women in business or politics. Her comment had nothing whatsoever to do with the gentleman's assertion that a strong man was needed as Secretary of Labor; but the little woman could think of nothing more disparaging to say at the moment, and she knew it would be an entirely acceptable remark!

There is one place, however, where sex must be a cleavage in daily activity. Women run their homes as women. They live their social lives as women, and they have a right to call upon man's chivalry and to use their wiles to make men do the things that make life's contacts pleasanter in these two spheres. Sex is a weapon and one that women have a right to use, because this is a part of life in which men and women live as men and women and complement, but do not compete with each other. They are both needed in the world of business and politics to bring their different points of view and different methods of doing things to the service of civilization as individuals, with no consideration of sex involved; but in the home and in social life they must emphasize the difference between the sexes because it adds to the flavor of life together.

Some people feel that the entry of women into industry has brought about the fact that there are not enough jobs. But we don't need to eliminate workers, we need to create jobs. We certainly haven't reached a point of satiety when all around us we can see that work needs to be done. Let us, therefore, as women unite for great fundamental causes; but let us insist on doing the work of the world as individuals when we wish to be modern versions of Mrs. Bloomer, and let us function only as women in our homes. We need not feel humiliated if we elect to do only this, for this was our first field of activity, and it will always remain our most important one.

We must be careful, however, to remain in the home not as parasites, but because our abilities lie along the lines of domestic life. Remember that a home requires all the tact and all the executive ability required in any business. The farmer's wife, for instance, must get into her day more work than does the average businessman. Many a woman runs the family home on a slender pay envelope by planning her budget and doing her buying along lines that would make many a failing business succeed.

It will always take all kinds of women to make up a world, and only now and then will they unite their interests. When they do, I think it is safe to say that something historically important will happen.

11

"You Can't Have a Career and Be a Good Wife"[1]

Ladies' Home Journal, January 1944

It is not startling to hear that Jennifer Jones Walker and Bob Walker have come to a parting of the ways. Always in double-career marriages there is terrific strain. With two strong personalities thrusting forward to success, not all the pull can be smoothly cooperative. There is crowding, and sometimes conflict. It could not be otherwise.

A marriage that survives twin careers is the exception; one that can *thrive* on a dual setup is a miracle. Many marriages, it is true, continue to be endured under these circumstances, for not everyone has the courage to face the situation as the Walkers are doing, admitting they no longer have a going concern, and taking steps to rectify matters. "Successful career couples," so-called, try to keep up a glossy surface, hoping it will not crack and expose the disappointing makeshifts underneath.

What is the matter with this dream of a man and a woman, both workers out in the world, making a solid home together? Why shouldn't two people, each with an outside job to do, unite all the more firmly to build a rich setting for their private lives? Why couldn't such a partnership be unusually full of understanding, and mutual respect, and lively ideas — twice as many of the last, in fact, as when only one partner goes afield to garner them? Why shouldn't the children of such progressive parents, with doubly wide horizons, be especially privileged and happy?

Unfortunately, it rarely works out that way. The picture of this ideal partnership has a deceptive and meretricious brightness. It is only after years of observation and experience that one realizes how superficial that

---

[1]The author of this article was identified only as "a successful career wife."

---

brightness is; realizes, too, that the best marriages do not necessarily glitter on the surface at all, but are solid affairs built on time-tested traditions. The reasons for this go very far back in human history, and have not been too much changed by household inventions or woman's suffrage or nursery schools or diaper services.

Most women want to keep on working after marriage for one of three reasons: 1. Because they want extra money; 2. Because they are lonely and want something to fill their days besides the care of a three-room apartment; 3. Because they feel they have something to express and have an honest urge toward work they really like.

Each of these reasons brings its own peculiar danger. The woman who works to supplement her husband's income is setting up financial habits which she will find increasingly hard to change. His money promotions, which would be important cause for celebration if he alone were the breadwinner, become less impressive. There is less sharp incentive to fight ahead. The double income becomes a pleasantly accepted fact. The baby they were going to have "someday" becomes a more and more distant prospect, involving too much rearrangement of their lives, too much sacrifice of comfort. After all, who wants to settle down to washing dishes—and baby things—when the extra money makes it possible to live more easily?

The wife who works to defeat loneliness is only postponing, or sidestepping in cowardly fashion, the period of adjustment necessary to make her new life successful. Marriage and home building should bring special interests of their own, not be simply an extension of the pattern of girlhood. The bride who continues to build her daily living around familiar office activities, friends, and gossip, instead of starting to develop additional friends and interests compatible with her changed condition, is as shortsighted as the young wife who wants to continue eating at mother's because she doesn't want to learn how to cook. Marriage requires rearrangement of many previous attitudes, and the best possible time to do that is right in the beginning.

Most insidious of all is the danger threatening the girl who really finds her work stimulating. A good job is a demanding job. The young woman who is giving her best to her work cannot give her husband and home all they deserve. As anyone who has fought forward in the world of commerce and careers knows, success comes from ideas, and ideas cannot be turned off and on by the clock. They arrive disturbingly at daybreak, or on Sunday afternoon, or when Aunt Minnie is coming to dinner. Often they call for an audience, or at least a pad, a stub of pencil and a few uninterrupted minutes.

Now every family is the better for *one* person who is throwing off ideas and creative steam in general. And "creative" in this instance applies just as much to schemes for selling plumbing supplies, or an improved system of bookkeeping, as to more colorful efforts in art and literature. It's good to have one partner vigorously emitting ideas and challenging the imagination with plans for the future. And that person needs an audience who is wholeheartedly interested, and who will put everything else aside to give him encouragement in critical moments. Because inside the heart of even the bravest fighter there is always a little loneliness and unsureness, and a great need of warm words. It is fine and sobering, too, for that fighter when he is out in front to know that the welfare of his family unit depends entirely on him and that he is backed by complete faith.

But when the front action is divided? When there are *two* fighters, each needing encouragement; two performers, each demanding an audience? What then?

If they could alternate their needs, that would help, but too often life, with devilish ingenuity, piles everything up at once. Caleb has a row with the boss and comes home to find Sylvia in tears because they have hired a new young woman in the department and given her the very work they promised Sylvia. Even worse is the New Year's when Caleb's raise doesn't come through on schedule and he comes wearily through the front door to be met by a joyful Sylvia shouting over an unexpected bonus. A man wants comfort and someone to share his grousing at the boss at a low moment like that, and no matter how many articles are written to prove it shouldn't be so, it hurts his male pride to have his woman winning, on her own, the business laurels he had hoped to lay at her feet. If this experience is repeated too often, he may become chronically embittered, or he may relax his own effort and become one of those subdued husbands often seen in the wake of successful businesswomen. In either case, the marriage is thrown off balance. An aggressive husband and an apologetic wife do not make a satisfactory couple, but it is even more against nature when the positions are reversed.

The problems of the children of a business couple deserve a whole volume. Presuming that a liberal financial setup, good health and flexible working hours for the mother have assured the arrival of a baby without too desperate a counting of time and pennies, the home life still takes on an incredible intricacy. Even the steadiest marriages, involving only a couple of adults, can be full of the unexpected; and when a baby—or more—is added to a household run by a working wife, things really begin to happen!

In spite of competent nurses, housekeepers and kindly relatives, there will be crises, arising from illness, minor accidents or the collapse of household personnel. In no time at all the career mother finds herself trying to be at least two people: an efficient, smartly turned out professional woman, and a devoted and conscientious mother. There is obviously little time left for being a wife.

There are inevitable penalties for this. Husbands become discontented as they feel themselves neglected; for no matter how much a business wife may be contributing to a mutual household in the way of an alert mind, an enriched personality and a wider circle of interests—quite apart from mere money—the husband who can't find his clean laundry considers himself abused and puts it all down to his unnatural home setup. It is of no use to point out to him that clean collars have been known to get mislaid by full-time wives. The real trouble, of course, is that in his heart, whether he himself knows it or not, he has an age-old male resentment of the fact that his woman is out in the world about her own business instead of staying safely in the cave he provides for her.

Conscious of her husband's feeling of not having an entirely normal home, his wife redoubles her efforts, tries to be a satisfactory mate as well as a harassed young mother and ambitious career girl. Something gives: nerves fray, health gives way. The tired businessman is merely suffering from mild fatigue compared with the tired business wife and mother. If her income is considerable, she can hire secretaries, cooks, nurses and maids, but the whole staff put together cannot take over the fundamental job of being the heart of her household, with the responsibilities of emotion and imagination which belong to a wife and mother. Was there ever a husband in the world who would prefer a hard-driven professional "success woman" to a relaxed, laughter-loving wife?

Yes, home wives work hard too. But they do not work at a stiff, competitive pace, with someone else calling the turn. They are their own directors. They are in business for themselves, and it doesn't take a psychologist to tell you that you can do twice the work, happily, under your own steam that you can be driven to do by outside forces.

Why, then, do women have this suicidal hunger for success outside the home? Why do they crave so deeply to "go out to business," instead of making a business of their homes? This does not apply to those women with rare abilities who will forge ahead and express themselves, and should, no matter what their environment. Most of the career-hungry girls would do much better as wives and mothers than as businesswomen. Arriving at middle age, and stopping to count the cost, they themselves wonder what they have attained that seemed so worth struggling for at twenty. Even the old argument that women in business stay more alert and hold youth longer is not true now—if it ever was. The modern happy wife and mother in her forties is apt to look like a better-adjusted and younger woman than her successful professional contemporary in thirty-dollar hat!

Who, then, is to blame for this exodus of women from the home? Mostly, *the men themselves.* The very men who would prefer to have their women stay by their own firesides have been extraordinarily inept in selling them the idea. And this in a nation of supersalesmen!

Take that heavy-handed, man-made phrase, "Woman's place is in the home," which men have kept repeating for generations. In that repetition, indeed, lies the trouble. It is not the lack of truth in the statement, but its terrible monotony. It sounds like a sentence, as though the woman—any woman—were going to be locked in a safe, and completely dull, place for life. You can almost hear the key turning in the lock. Any woman with spunk resents this attitude. Why should any man, even her beloved husband, assume the right to shackle her to a spot, or to a set of obligations? Even though he promises to treat her kindly, if she pleases him!

Oh, what a poor selling job the males have done all these years! They, whose selling ability has become legendary, have fallen down on the biggest job of all, the job of making the little woman feel that she is so extraordinarily lucky just to be a woman, and a potential wife and mother, that she really ought to hug the knowledge to her heart in a kind of secret joy lest she be accused of vulgar boasting.

Instead, the obtuse male sets about destroying his own future peace of mind, as far as marriage is concerned, while he is still short-trousered. The first time he growls patronizingly to a feminine schoolmate, "Aw, you're just a girl. Whadda *you* know anyway?" he is planting seeds of discontent. At that instant there may be born in his pigtailed mate a determination to "show him someday." The *why* of that determination will vanish, but the feeling of resenting male belittlement may be carried on and translated into powerful activity ten years later.

There is no perfect solution, of course, but if the men who are complaining about what seems to them a topsy-turvy arrangement today, and the younger men who are worrying about an even greater invasion of women in business tomorrow, will *stop scolding and start selling*—well, women have always been a responsive buying audience!

In the meantime, in spite of the serious handicaps involved, some couples will make a good job of double-career marriages by being willing to work extra hard and to learn the great law of compromise. But it takes an exceptional woman—and an even more exceptional husband.

12

ALICE HAMILTON, M.D.[1]

"Why I Am against the Equal Rights Amendment"

Ladies' Home Journal, July 1945

We hear a good deal nowadays about the Equal Rights Amendment, which, after more than twenty years of ardent effort on the part of a group of feminists, the Woman's Party, seems to have reached a point where its passage by Congress is possible. Both parties in the last election declared themselves in favor of it, and it has been passed on favorably by the Senate Judiciary Committee.

What is this amendment to the Constitution which may be put up to the state legislatures in the near future? It reads very simply:

"Equality of rights under the law shall not be denied or abridged by the United States or by any State on account of sex. Congress and the several States shall have power, within their respective jurisdictions, to enforce this article by appropriate legislation."

Well, that sounds only fair and reasonable. No woman wants to have her rights denied or abridged. The amendment seems to open a closed door, to clear away difficulties that have been in her path. Perhaps she is working in an office where some man, not nearly so efficient and conscientious as she, has a higher place, just because he is a man. Perhaps she is a physician who, because she is a woman, cannot get on the staff of a hospital. Perhaps she is a teacher who knows that when the principal retires she would get his place if she were a man, but no matter how much she deserves it she will be by-passed and a man put in. Or perhaps she is a businesswoman like one who said to me, "Women in business are like the Red Queen in Alice; we have to run as hard as we can to keep in the same place."

---

[1]Alice Hamilton was president of the National Consumers League, which was founded in 1899 to foster citizen participation in government and industrial decision making regarding such matters as fair labor standards and product safety. The league still exists, with headquarters in Washington, D.C.

Alice Hamilton, "Why I Am *against* the Equal Rights Amendment," *Ladies' Home Journal,* July 1945, 23.

These women see in the amendment a means of wiping away all such sex discriminations; they see themselves placed on a real equality with men. But are these hopes justified? Will such a law do that? Will it give women more than it will take away from them? I do not think so, and I will tell you why.

No law can compel a man to employ a woman or to promote her, no law can compel a hospital to place women doctors on its staff or to admit them as interns and residents, no law can prevent an employer from passing by a competent woman and appointing a less competent man. These are matters which lie outside the domain of law; they are decided by men who are often swayed by the old prejudices, the instinctive inhibitions and compulsions which centuries have implanted in us. I do not mean that these obstacles cannot be overcome; the history of the past century shows how far the emancipation of women has advanced already, but the advance cannot be brought about by law, by compulsion. Only the slow growth of a genuine feeling of sex equality can bring it about; but surely when we see the attitude of the young generation we cannot fail to be heartened by the striking signs of increasing equality between the sexes.

It is true, as the advocates of the Equal Rights Amendment insist, that there are still in some states inequalities that can be removed by legal action, and have been in the more advanced states. To that I would answer that it is in the power of such states to follow the example of their sister states, that it is not necessary to bring about such relatively simple reforms by so cumbersome a method as an amendment to the Constitution.

But why should we—I speak as president of the National Consumers League and as a member of the National Women's Trade Union League, the League of Women Voters, the Association of University Women— why should we who call ourselves feminists oppose a measure which, even if it does not do all its proponents claim, will still help to remove women's disabilities? It is because we know that women would lose more than they would gain if at one stroke all the laws which treat women as women and not only as persons were revoked. Now to the strictly orthodox feminist, women should be treated only as persons. I remember a debate before a women's club in which I took the side opposed to the amendment and the side defending it was taken by an eager, attractive but very young woman. I argued that the laws, so general in the older countries, which provide vacation with pay for pregnant women in industry were necessary for the good of the race. She opposed it, of course, but finally gave way. "All right," she said impatiently, "if you say 'pregnant persons' we won't object, but not 'women.' " I assured her we would be quite content with that.

This is, of course, an extreme instance of what one might call "feminist fundamentalism," and I am not suggesting that it is typical of the Woman's Party. But I do think it is typical of that group to ignore the real physical and social differences between men and women, differences which lie at the base of the laws applying only to women and not to men. The fact that these differences have often been exaggerated in the past and that nowadays we ignore many of them does not mean that we can sweep them all away and act as if that settled it. Women still have to bear the babies and still ought to rear the children, and I ask any young mother if that is not a real handicap which deserves consideration.

The proposed amendment would do away with the protection of women as homemakers and mothers; it would either make the wife equally responsible with her husband for the support of the family or it would remove responsibility from both. In such a case, which parent would be more likely to get out from under? Ask any social worker. Surely it is only common sense to hold the husband responsible for the support of his family.

And what about alimony after divorce? So far the laws relating to alimony (passed by male legislators!) have been much more favorable to women than to men; in fact, they have shown an absurdly sentimental attitude and often they have worked great injustice. But to abolish them altogether and to put the woman on an equal footing with the man would work even more injustice. Take the middle-aged woman who has no professional or industrial experience, who knows no skill but amateur housekeeping and child care. When her husband tires of her and seeks a younger mate, is she to be left penniless to find her own support? Alimony laws need strict revision, but not abolition.

The laws of property as they affect married women are too complicated to be covered in this discussion. It is quite true that in backward states such laws are unfair to women, but in most of the states they have changed with the changing times and those disabilities that still remain can be removed if ardent feminists will only carry on the sort of intensive campaign which we of the older generation devoted ourselves to, some forty to fifty years ago. They would then know exactly what they were trying to achieve, for they would be attacking a definite evil, not a system which is part evil and part good.

It is in connection with labor legislation that my opposition to the Equal Rights Amendment is strongest, for my field for thirty-five years has been the protection of working people against the harmful features of industrial work, against dangerous dusts, poisons, excessive heat and heavy exertion, long hours, low wages. Dangerous dusts cause silicosis, followed sometimes by tuberculosis; poisons used in industry include lead, mercury, mineral acids, carbon monoxide, and a host of volatile solvents which act like ether and chloroform, only sometimes more harmfully. Our efforts have always been to protect both men and women against these dangers, but when we could not do that, when legislators were willing to protect only the women, then we took gladly what we could get.

And in some respects protection is needed more by the women. For one thing, they have the handicap of youth; and, as we all know, the younger the worker, the more susceptible to a poison and to fatigue and to the harmful effects of dangerous dusts. Men in industry are divided into the different age groups just as men are in nonindustrial work, but women in industry are massed overwhelmingly in the early age groups. This means that they are more susceptible to all the harmful elements in their work, they are less prudent than older women, they are reckless and they have the silliest ideas about food and about protection against cold and wet. Partly because they are young, partly because they draw the lowest wages, young women workers have a higher tuberculosis rate than men. Young women workers cannot stand for long hours on their feet as well as men can, nor can they lift heavy weights; older women have their special handicap, housework and the care of children added to the hours of factory work.

It is true that the Walsh-Healy Wage and Hour Law[2] applies to men as well as women, but it covers only those industries that enter into interstate trade, not those that are strictly within the state — restaurants, hotels, laundries, stores and the white-collar industries. For these, state laws are necessary, to prescribe a minimum wage and maximum hours. Of course it would be better to extend such laws to cover men also, but this has never been done. We should be giving up a reality for a vain hope.

We do not need, we older women, to ask what will happen if all the laws protecting workingwomen should be revoked, for we remember only too well what happened before those laws were passed. In Illinois we fought for years for a ten-hour law for women.[3] I knew a young Irish

[2]The Walsh-Healy Wage and Hour Law (1938) established the first minimum wage (25 cents an hour), mandated time-and-a-half pay for working more than 40 hours a week, and prohibited the employment of children under 16.

[3]Ten-hour laws, passed in many states and upheld by the Supreme Court in the early twentieth century, usually prescribed the maximum hours women could work per day (ten in some states, twelve in others). Designed to provide protection for women not accorded to men, the laws caused divisions in the women's movement throughout the century until they were overturned by federal legislation in the 1970s.

girl in Chicago who had a job as night waitress in an all-night eating place out near the stockyards. She worked six nights a week, twelve hours a night, seventy-two hours a week, and it could have been raised to eighty-four so far as the law was concerned. There were no minimum-wage laws in 1912 when, out of five million workingwomen in the United States, half under twenty-four years of age, a million earned less than $4.00 a week.

In those days I hated to shop in Chicago stores, for I knew so well the lot of the girls who worked in them. There was no vacation; not even in the hot summer months could the employers be persuaded to close for one weekday. It was a ten-hour day with plenty of overtime—paid for usually by a lunch—especially before Christmas; wages ran from $2.50 to $10.00 a week and very few ever reached that latter figure, yet a girl could not then support herself in Chicago on less than $8.00 a week. Even to get seats for these girls meant a long fight. I remember a girl who was a worn-out, middle-aged woman at twenty-four, after ten years' work in a department store for never more than $7.00 a week, all of which went to her widowed mother. She had had no normal girlhood, no fun ever, for her Sundays must be spent in sleep or in mending and laundering.

The women who are pushing the Equal Rights Amendment insist that to limit the hours of work for women and to insist on pauses for lunch, on separate toilets, on provision of seats—all these things handicap women in getting jobs, and that they want to be free to compete with men on equal terms. I think that is probably true of a small number of women in industry, such as linotypists on morning papers, and ticket sellers on elevated roads and subways, who are not allowed to work on the night shift. But in normal times their number is infinitesimal compared with the number of women who welcome these safeguards. Of course in wartime all restrictions are suspended in most of the states.

I can assure you that when Ohio passed laws for the protection of women workers, Ohio women did not envy the freedom of the women of Kentucky to work as long hours and for as low a wage as they pleased; Massachusetts textile workers did not move to Rhode Island to free themselves from the wages-and-hours laws; the women of the Northern States did not look with longing at the untrammeled freedom of the South. There is no articulate organization of workingwomen that favors the Equal Rights Amendment.

For more than half a century the struggle to alleviate the lot of women workers has gone on, and we can point with pride to what has been gained. Twenty-six states have minimum-wage laws, and this means that some four million women in hotels, restaurants, laundries, and so on, must be paid a decent wage. Forty-three states limit the hours which a woman may be required to work. Less essential, but still important, are the regulations which require seats—and their use—separate toilets, time off for lunch, and so on, and these are in force in practically all states. And such laws have not limited the employment of women, handicapped them in their search for jobs; on the contrary, it is during the past twenty-five years, when these laws were passed, that the employment of women has increased steadily—not only during the war—and in just those states where labor laws are strict. All this gain would be swept away by the adoption of the Equal Rights Amendment. Is it any wonder that workingwomen have dubbed it the "Unequal Rights Amendment"?

In theory, of course, women should fight for their rights as men do, but in actual fact they seldom do. Even now, in wartime, only about three million of the seventeen and a half million women wage earners belong to unions, and the majority of these are in war industries, aircraft and automotive production. When the war is over few of them will remain in those industries; they will have to go back to their old jobs, most of them in unorganized industries or with far less strong unions. Surely they will need as much protection as they did before the war.

The Equal Rights Amendment will not free women from discrimination; no laws can do that. It will not make it easier for professional women to advance, nor for workingwomen to get or to hold jobs, but it will sweep away the few measures which now make the lives of the wage-earning women less hard. It will remove from the wife and mother all consideration for her as a woman and will place her on a legal equality with her husband, an equality that is mostly imaginary, has little basis in hard fact. The legal injustices under which women in some states still suffer can be removed one by one with far less effort than was needed in earlier years, and without sacrificing what is good in order to get rid of what is bad. If women in and out of industry understood clearly what the proposed amendment would do, it could never be passed.

13

JENNIFER COLTON
"Why I Quit Working"
Good Housekeeping, September 1951

Just over a year ago, I was suffering from that feeling of guilt and despondency familiar to most working mothers who have small children. During the hours I spent in the office, an accusing voice chanted continuously, "You should be home with the children." I couldn't have agreed more, which only created an additional tension: the frustrated anger of one who knows what is right but sees no way of doing it. Children need clothes as well as attention; they must be nourished with food as well as love.

Some mothers can eventually talk themselves out of this feeling; they conclude that the material advantages they can provide are worth more than their presence in the home. Some, unable to rationalize this way, gradually grow more despondent. Others try to find a compromise. For the fortunate ones, the problem is solved by a substantial increase in the husband's income.

One day in 1950, I finally worked out a compromise: a way to be at home with the children and still do some work for which I'd be paid. At that moment, I knew only two things: that I would never again rage against a delayed subway for costing me my painfully brief hour with the children, and that at last I would be able to serve a dinner that took more than half an hour to prepare and put on the table.

A year has passed, and I've had time to judge the advantages and disadvantages of leaving my office job. How they will total up ten years from now, I don't know. But here is my balance sheet of the results to date.

Lost

The great alibi: work. My job, and the demands it made on me, were my always accepted excuses for everything and anything: for spoiled children, neglected husband, mediocre food; for being late, tired, preoccupied, conversationally limited, bored, and boring.

The weekly check. And with that went many extravagances and self-

---

Jennifer Colton, "Why I Quit Working," *Good Housekeeping,* September 1951, 53.

indulgences. I no longer had the pleasure of giving showy gifts (the huge doll, the monogrammed pajamas) and the luxury of saying "My treat." And without the extra money, I couldn't rectify or camouflage such mistakes as an unbecoming hat, a too-big canasta debt, a too-small pair of shoes. *The special camaraderie and the common language.* The warm but impersonal and unprying relationship among working people is one of the most rewarding things about having a job. People who work, even in unallied fields, speak rather the same language, which can't be translated for the uninitiated without going into the whole psychology of business. I missed the crutch of shoptalk when, later, I struggled to reach people through interests outside the business world.

One pretty fallacy. For some reason, most working mothers seem to think they could retire with perfect ease; that they could readily adjust themselves to their new role. I don't think so. When you start to devote all your time to homemaking, you run into a whole *new* set of problems. The transition from part-time to full-time mother is difficult to make.

One baseless vanity. I realize now (and still blush over it) that during my working days I felt that my ability to earn was an additional flower in my wreath of accomplishments. Unconsciously—and sometimes con-sciously—I thought how nice it was for my husband to have a wife who could *also* bring in money. But one day I realized that my office job was only a substitution for the real job I'd been "hired" for: that of being purely a wife and mother.

The sense of personal achievement. A working woman is someone in her own right, doing work that disinterested parties consider valuable enough to pay for. The satisfactions of housekeeping are many, but they are not quite the same.

The discipline of an office. The demands made on you by business are much easier to fulfill than the demands you make on yourself. Self-discipline is hard to achieve.

Praise for a good piece of work. No one can expect her husband to tell her how beautifully clean she keeps the house or how well she makes the beds. And other people take her housewifely arts for granted. But a business coup attracts attention.

Found

A role. At first I found it hard to believe that being a woman is some-thing in itself. I had always felt that a woman had to do something more than manage a household to prove her worth. Later, when I understood the role better, it took on unexpected glamour. Though I still wince a little at the phrase "wife and mother," I feel quite sure that these words soon will sound as satisfying to me as "actress" or "buyer" or "secretary" or "president."

New friends and a wider conversational range. It was sad to drift apart from my office colleagues, but their hours and, alas, their interests were now different from mine. So I began to make friends with people whose problems, hours, and responsibilities were the same as mine. I gratefully record that my friendship with them is even deeper than it was with business associates. Although we share the same pattern of life, we are not bound by it. As for conversation, I had been brought up on the satirical tales of the housewife who bored her husband with tiresome narratives about the grocer and the broken stove. Maybe it was true in those days. But not any more. I've had to exercise my mind to keep up with these new friends of mine. They have presented me with that handsome gift of expanded interests.

Normalcy. The psychiatrists say there is no such thing, but that's what it *feels* like. My relationship with my children is sounder, for instance. I have fewer illusions about them. I have found I can get bored with them. Exhausted by them. Irritated to the point of sharp words. At first I was shocked, and then I realized that when I worked and we had so little time together, we had all played our "Sunday best." The result: strained behavior and no real knowledge of one another. Now I'm not so interesting to them as I was. I'm not so attentive and full of fun, because I'm myself. I scold, I snap, I listen when I have time. I laugh, I praise, I read to them when I have time. In fact, I'm giving a pretty good representation of a human being, and as the children are going to spend most of their lives trying to get along with human beings, they might as well learn right now that people's behavior is variable.

The luxury of free time. This is one of the crown jewels of retirement. The morning or afternoon that occasionally stretches before me, happily blank, to be filled with a visit to a museum or a movie, a chat with a friend, an unscheduled visit to the zoo with the children, the production of the elaborate dish I'd always meant to try, or simply doing nothing, is a great boon.

Leisure. The pleasure of dawdling over a second cup of coffee in the morning can be understood only by those who have, sometime in their lives, gulped the first cup, seized gloves and bag, and rushed out of the house to go to work.

Handwork. This may seem trivial, but making things at home is one of the pleasures the businesswoman is usually deprived of. Homemade cookies, presents, dresses, parties, and relationships can be worth their weight in gold.

Intimacy. The discovery of unusual and unexpected facets in the imaginations of children, which rarely reveal themselves in brief, tense sessions, is very rewarding.

Improved Appearance. Shinier hair, nicer hands, better manicures, are the products of those chance twenty-minute free periods that turn up in the busiest days of women who don't go to business. Of course, such periods crop up in an office, too; but you're not allowed to make use of them for personal affairs.

Proof Positive. If I hadn't retired, I would have remained forever in that thicket of self-delusion called thwarted potentials. It was almost too easy: the shrug, the brave little smile, and the words "Of course, I've always *wanted* to write (or paint or run for Congress), but since I'm *working,* I never have time." And it's time that gives you proof positive of what you can and cannot do.

Relaxation. Slowly, I'm learning to forget the meaning of the word tension. While I was working, I was tense from the moment I woke up in the morning until I fell into bed at night.

There is no way of measuring or comparing unrelated work. I don't know whether I work harder or less hard now. I walk farther, but there are often free periods during the day to enjoy as I like. I do a greater variety of things, but at my own speed, and without the pressure common to all offices. I get sleepy instead of tired.

Sometimes I ask myself, "What would persuade me to go back?" And my answer is, "Barring big medical expenses or a *real* need for something for the children or my husband, nothing." And I mean it.

But I'm glad I had the experience of working. I can understand my husband's delight in his work, and I can still talk sympathetically with friends who work. And what else could make me so acutely conscious of every blessing and so humbly aware of the potentials of my new role?

14

"Women in Flight"

Mademoiselle, December 1952

"Make up your mind to stick on the job six months. It may take you that long to find out you really enjoy it." Experienced airline hostesses often give this tip to newcomers. It's based on the down-to-earth fact that being welcomer, waitress, comforter, housekeeper and sage to all sorts of people at all sorts of altitudes is something you have to get used to. Most hostesses feel that the adventure and fun of their work—and such assets as the prestige of their uniforms—outweigh the rough parts.

But the rough parts are there. A hostess may leave home base at 3:00 A.M. (perhaps she has been on call for hours or even days before); she may live in three climates during her trip, then land at her destination in an after-work mood. The rest of the world will be moving in its mid-morning routine. Until a hostess can take topsy-turvy hours in stride they may thoroughly blunt the pleasure of meeting a hundred new people a day, of being a commuter between States.

On paper the working hours seem utopian. Total flying time for a month is set at eighty-five hours—which averages a tidy twenty-one-hour work week. The airlines calculate that while in flight one hour a hostess uses as much energy as she would in two hours on the ground.

Likely a girl would not apply for a hostess job—she certainly wouldn't be accepted—if she didn't have great skill and joy in dealing with people. This helps her to smile with sincerity at a wet witticism she's hearing for the thousandth time. She must be equally pleasant about ministering to the airsick and reassuring fretting passengers.

A hostess, prescribes one airline, should have charm, personality, poise and beauty. Although in practice most companies dilute this redundant demand for perfection, a girl who lands a hostess job has survived a weeding-out process. The chief instructor of American Airlines' stewardess school says: "To get a class of forty trainees we may interview two thousand girls."

To qualify for any airline, a hostess must be single (she's asked to resign when she marries) and she must pass a rigid physical examination. She cannot wear glasses—or need them—on the job. Height and weight requirements vary slightly, but the average is 5′2″ to 5′7″, 105 to 135 pounds, with weight pleasantly distributed. Delta Air Lines accepts hostesses of 5′ and Northwest goes up to 5′8″. Age requirement averages from twenty-one to twenty-eight, with Northeast accepting girls of twenty. College, college plus experience working with people or an R.N. are usual requirements.

For duty on international flights a hostess usually needs from one to three years of experience on domestic runs and fluency in a language. Pan American, entirely an international operation, hires for immediate duty abroad and has no language requirement for its Pacific-Alaska division.

When a girl has forded successfully a series of interviews and tests she is usually given four to six weeks' training, with room and board plus a salary of $175 in some cases. All large airlines have their own schools, although TWA, Eastern and some of the small companies also recruit from private schools.

While training, a hostess learns the answers to the questions passengers will be asking her in the future—from how pilots fly on radio beams to why clouds drift into a particular formation. She practices serving lap lunches and goes through many drills on how to act in a forced landing or crash. Pay usually starts at $190, plus maintenance away from home. Salaries can go to $285.

15

"The Married Woman Goes Back to Work"

Woman's Home Companion, October 1956[1]

Women are going back to work—a vast, new army of them. That's the big news in industry today.

Woman's hard-won right to work outside the home has developed into a powerful urge to work. Between now and the first of the year many more

---

[1]The following statistics accompanied the text of this article: Of the 66 million Americans working outside the home in 1956, one-third were women; of these 22 million women in paid employment, 6 million were single and the rest were married, and more than half of the married women had children of preschool or school age. The six types of jobs identified were populated by women as follows: clerical 6,600,000; factory 5,500,000; service and domestic 5,100,000; professional 2,100,000; sales 1,600,000; and farm 1,100,000.

"The Married Woman Goes Back to Work," *Woman's Home Companion,* October 1956, 42.

women will enter the business world, emerging with money, prestige and new friends. Their drive to work is exceeded only by their employers' delight in having them aboard. For the woman worker is being courted with unprecedented ardor by big business.

The odds are high that you will be going back to work (one in three women do) and that you have questions about this new way of life. This article, prepared with the help of the Crowell-Collier Regional Offices, the National Manpower Council and the Department of Labor, is designed to answer your questions.

What kind of women will you be working with? Most of the millions who will be added to the nation's present working force of 22,000,000 women (an all-time high) will not be youngsters fresh from school. On the contrary, the majority will be mature housewives. The typical new woman worker will be at least thirty-five years old, married, with children.

For several years now young girls have continued to go to work at the same rate as before, but older women account for the big upsurge in new employees. Many have never worked before. Others have not held jobs in ten or twenty years, the years in which they bore and raised children.

Today, for example, almost half of all the women in the country between the ages of forty-five and fifty-four are working. And they are not primarily widows, divorcees or spinsters. They are mothers and home-makers.

Why are women rushing back to factories and offices when they could be their own bosses at home? What is the lure that may persuade millions more, including you, to join the ever-swelling working ranks and make the stay-at-home wife the exception, rather than the rule?

The answer seems to be threefold.

First of all, there's the obvious answer—money. But, statistics show few women need the money for survival. As the personnel manager of a large Texas plant put it: "So far as we can learn, few of our women really *have* to work."

Or, as the psychiatrist of one of America's largest industries summed it up: "Married women work to meet the cost of high living not the high cost of living!"

A woman in a Jacksonville cigar factory explained, "We could exist on what my husband makes but we wouldn't be as happy."

The "extras" a working wife brings into the home soon become part of a fixed standard of living and become a habit. The longer a wife works, the more difficult it becomes to separate necessities from "necessary" luxuries.

"If we can keep a woman on the job for a year," the personnel manager of a large corporation confided recently, "we find we can keep her as long as we want to. Her family gets so used to the extra income that they can't imagine doing without it again."

The second thing that drives women out of the house and into an office or factory to earn a weekly pay check is prestige.

Women, no less than men, need to feel successful. Success as a house-wife is not easily measured.

Money, on the other hand, is often the main measure of success in the "outside world." A job puts a definite, tangible value on time, value that is measured in dollars.

Some experts have said that modern laborsaving devices make house-work too easy to keep women busy, too mechanical to be satisfying. A busy mother knows better. She knows that laborsaving devices do not entertain the children and their friends or pick up after them or run those extra errands.

Contrary to opinions voiced by "experts" (most of them men), today's wife does not go to work because she has nothing to do at home. She goes to work because what she could be doing at home is not highly valued by society (including herself).

Nor does she leave her house because she dislikes the nature of housework. She probably will have to do the same housework in any event; her schedule will be different but her chores will remain the same.

Housework can be repetitious, monotonous, mindless. Perhaps much of it is. But if the average housewife agrees with that analysis, she is jumping from the frying pan into the fire when she takes an outside job. Most of the married women in the new labor force are to be found in two fields that are notorious for their monotony and repetition—factory work and clerical work.

Yet women from one end of America to the other claim that going back to work is "stimulating." If the jobs themselves are not stimulating, what is?

"There's a nice gang here," a factory worker said recently on her ten-minute break during a morning of putting screws into the backs of radio sets. "We get together at lunch time in the company cafeteria. It sure beats fishing a leftover out of the icebox all by myself."

"At last," a woman, who operates a punch machine in a printing plant, said, "I have something to talk to my husband about in the evening. Look"—she gestured toward the near-acre of machines and people— "things are really going on here."

An articulate billing clerk, who left college to marry when she was 20 and is now holding her first job at 37, remarked, "I was lonesome at home. I've rejoined the human race and it's wonderful! See, here's an order from Buenos Aires and the other day I had one from Morocco. When I get home at night I feel as if I'd really been somewhere." She smiled. "And you should see how impressed my children are!"

One of the pleasant factors about going back to work today is that women are welcomed with open arms. There was a time, back in the thirties, when married women were fired, not hired. With husbands unemployed, no one wanted to hire wives. But today there are more jobs than people to fill them. The national shortage of teachers, nurses and scientific workers is well known. And jobs that require little or no special training are going begging too. According to the Women's Bureau of the U.S. Department of Labor, at least 600,000 clerical workers are now needed in business.

There is still, however, one formidable obstacle faced by older women in job-hunting and that is the pension plan. Companies with elaborate retirement benefits are highly sensitive to the age question. They prefer to hire younger workers so they will have time to accumulate a pension fund.

Group life insurance is another factor that presents an obstacle. This kind of insurance costs the boss more when the average age of his employees goes up.

Despite these disadvantages, however, companies are finding ways and means to add older women to their payrolls. One of the schemes involves getting workers from a new kind of service bureau that specializes in "temporary" employment. These new bureaus, of which Manpower Inc. and the Russell Kelly Office Service represent the largest on

---

As life expectancy increases and technological improvement decreases work in the home, more older women will inevitably join the labor force. There is no reason to fear this trend. The challenge lies in learning how to help some employers to adapt themselves to older women, and vice versa. More careful studies of problems involved are needed. Early success in these efforts means happier and healthier families.

DR. LEONA BAUMGARTNER
Health Commissioner of New York City

a national scale, send the employer as many workers as he wants for as long as he wants them with no strings attached. The workers are employed by the service bureau and so the client company has no pension responsibilities to them.

Manpower Inc. is so hungry for woman-power that it gives its present workers dishes and other premiums for sending in their neighbors and friends. To persuade the woman to take more jobs, more days of the week, Manpower Inc. offers both prizes for and pointers on household efficiency.

"It almost seems as though we can sell all the time that all the housewives in the country can give us," Elmer Winter, the president and cofounder of this particular service bureau, said recently.

Perhaps you are unable to quite make up your mind whether or not to become a working wife. Perhaps your household can now or will soon be able to spare you. Before you decide to jump onto the back-to-work bandwagon, however, you would be wise to take a long, hard look at yourself—your domestic setup, your job qualifications and your own makeup.

Here are the important questions to ask yourself:

1. Can you afford to work?

This is no joke. It costs more to work than to stay at home. Unless you can afford to think of a job as a mere pastime, a form of self-improvement or training for the future, you will probably want to break even at least. How much will you have to earn to pay your job expenses?

Your additional personal expenses will probably include a more extensive (or more expensive) wardrobe than you need at home, the cost of maintaining that wardrobe (dry cleaning, shoe repair, nylon replacement), beauty care (no time to do it yourself), transportation costs, contributions to office collections, club or union dues, telephone calls and

---

In years gone by the married woman of forty or fifty was ready for the shelf. Today modern miracles are allowing women not only to live longer but to retain their youth as well. The time of feeling useless is past for all women, regardless of age.

IVY BAKER PRIEST
Treasurer of the United States

lunch hours (both what you eat and what you buy when you're window-shopping from twelve to one).

Additional household expenses can be substantial. When pressed for time, you may send more laundry out, get more housecleaning help in. You may switch to a dry cleaner who delivers (and charges more), have your milk delivered instead of picking it up yourself. You may fall into the expensive habit of ordering your groceries by telephone or buying at a fancy food shop because it's handier or open later than your old reliable supermarket. You may resort to a whole string of kitchen short cuts, from cooking short-order chops instead of long-cooking economy meats, to paying bakery prices for cakes you used to make yourself. If you have a sitter or a maid you'll have to figure in the cost of her meals.

But then, of course, if you have a sitter or a maid, you are already committed to earning about half again as much as you pay her, simply to break even. The crucial facts are these:

A. In most cases you cannot deduct from your income tax return any part of what you pay out to sitters or housekeepers. The government does not consider the cost of your replacement at home a "business expense," even if your situation is such that you cannot work without a replacement.

B. Whatever you earn will be taxed *on top of your husband's income,* in the highest tax bracket his income reaches. The more money he makes, the higher percentage of your earnings will go for taxes. He is already, no doubt, filing a joint return in which he takes all the deductions you are both allowed and divides his income with you.

If you add income without adding deductions, your earnings must therefore be figured at the top bracket. If the last $1,000 of his income is taxed at the rate of 30 percent, say, then for every dollar you earn you'll have to set aside 30 cents for taxes.

Our free way of life is a constant example and a challenge to other nations. The status of the American woman is of particular interest to millions of women throughout the world. By treating women workers as individuals, judging them by their capacities rather than by their age, American business—and the entire free enterprise system—is proving its fairness, its adaptability and its enormous productivity.

MARY PILLSBURY LORD
United States Representative
United Nations Human Rights Commission

Even if you make only $600 during the year in such a situation, you will have added $180 to your family's income tax bill. This may be perfectly all right, particularly if it doesn't cost you very much to earn the $600. But you had better face this tax bill before you run it up. A mother of three didn't. She took a job as an assistant nursery school teacher, paying 75 cents an hour. She found a sitter who would look after her youngest child for 75 cents an hour. She didn't expect to make a profit, of course, but she did expect to break even while enjoying a stimulating work experience. At income tax time she discovered that her job was costing her 23 cents an hour for federal income tax alone. She has decided she can't afford to work next term.

To be absolutely realistic, then, you should add up all the hidden costs that may be entailed in your working, then figure out how much you will have to earn in order to pay these costs plus your income tax. If your expenses will be $50 a week, and your husband's income will put your whole paycheck into the 30 percent bracket, you'll need a salary of $71.50 to break even.

Such arithmetic will tell you what kind of job you can afford. But what kind can you get? That's the next question . . . .

2. *What can you do?*

Your home and community experience may have given you a certain balance, maturity, organizational ability and warmth, but if that's all you have to sell, your employer is not likely to buy. The years you have devoted to the homemaking arts will turn up as a large blank space under "Experience" on the application.

That, and not your age, is the major handicap in today's labor market.

If several years will elapse before you can possibly consider a job — if, that is, you are presently in the most demanding stage of your life as a wife and mother — you may actually be in a better position than the woman who wants a job right away. You have time to plan your return to work (a sound idea even if you think you'll never need to support yourself).

Teaching, for example, may strike you as a likely field. If you have a college degree, you may be eligible for an accelerated course that will prepare you for elementary school teaching. Such a course is being given at Wayne University in Detroit. For $220, plus eighteen weeks of her time, a college graduate between twenty-one and fifty can earn a five-year teaching certificate and a place in Michigan's elementary school system. Her course includes practice teaching in a school near her home plus for-

> Abilities and attitudes, not anniversaries, are the criteria in employing women today. Mature women in our labor force now number one of every eight workers and contribute significantly to the maintenance of our country's high-level economy. Planning specific programs to assist them is one of the principal interests of the United States Department of Labor.
>
> ALICE K. LEOPOLD
> *Assistant to the Secretary of Labor*
> *for Women's Affairs*

mal classes in the closest of three Detroit-area "workshops." She walks into a job that pays around $4,000 a year, with summers off, and she usually turns out to be at least as good at her work as the graduates of traditional teachers' colleges.

In the nursing field you can study to be a practical nurse if you are under fifty and have had two years of high school. You might also learn hospital administration with the hope of relieving registered nurses of some of the administrative work that now bogs them down.

Science needs thousands of qualified researchers and Dr. A. B. Kinzel, vice president in charge of research for the Union Carbide and Carbon Corporation, says that women are no less welcome than men. However, considerable training is needed in that field. Nevertheless, if you have a B.S. degree and a natural bent in this direction, you might well look into the ever-growing need for science workers.

Offices, of course, still absorb large numbers of the women going to work—filing, stenography and electric-machine punching. And the want ads will show you that a good secretary is forever hard to find. Older women are particularly welcome in small offices where a sense of responsibility counts for more than ravishing looks. There they are expected to keep books, handle the switchboard and make themselves generally useful.

As you see, jobs are waiting but they call for qualified workers. Few people are yet prepared to take you by the hand and lead you to the training you need to qualify yourself. That's up to your initiative.

3. *What does your husband think of you working?*

If your husband thinks it is a major crisis for you to go off to work when your child seems to be getting the sniffles (or when the lilacs need prun-

> Work is essential to a happy life. The older woman who finds that family responsibilities leave her time for outside work should look for a job. It may be paid or volunteer but should give her a chance to use her skills and earn the satisfaction of being needed and appreciated.
>
> LILLIAN GILBRETH
> *President of Gilbreth, Inc.*

ing or his client's wife needs entertaining), the pull to stay home will be greater than the push to live up to your commitments as a working woman. But, if he thinks your job is a sound idea basically, and if he works with instead of against you on whatever adjustments you have to make at home, you'll be spared a lot of agonizing self-doubt.

4. Are you healthy and energetic?

Can you cook and wash dishes and sort laundry and be reasonably entertaining after the proverbial "hard day at the office"? A working wife has two jobs, so naturally she needs more "git-up-and-go" than either a stay-at-home wife or a working girl without household responsibilities.

You may find that a job actually recharges you, that you have more energy in your evening homework because you have done something different through the day.

But it can work the other way too. If your present schedule leaves you limp at the end of the day, the cause may be something more than boredom. Taking an outside job won't noticeably diminish your housework: you can count on that. So you had better be sure you can take it physically before you go back to work.

5. Are you adaptable?

Many employers believe that older women are "set in their ways," that housewives are too independent to work under supervision, that mothers tend to be bossy. You won't be, at least not if you are aware of the possibility and take extra pains to behave like the novice you are in the workaday world. Your biggest problem in this area may be to forget your own age. If you can do that, chances are you won't be sensitive to criticism from that "mere youngster" who happens to be your supervisor. If your age doesn't matter, then neither does his.

Being adaptable at home may be somewhat more difficult. Will you wear yourself to a frazzle trying to live up to your former standards of housekeeping? If you cannot take short cuts and "let things slide" without feeling guilty or frustrated, it may be that you are not adaptable enough to be a working wife. And you'll probably never know until you try.

In making up your mind, however, this thought may help. Whatever you decide about going to work need not be your last word on the subject. If you decide to prepare yourself for a job, you don't necessarily have to take one. And if you do take one, you don't necessarily have to stick with it for the rest of your days.

And remember, too, that it is never too soon to plan your entry into the labor force. The younger you are when you go back to work, the greater experience you gather now while jobs are easy to get. Also, the better equipped you will be to survive any tightening that may later occur in the job market. Working now could add a great many plus values to your life, not the least of which would be the knowledge that you could support yourself and your children if you ever really had to.

3

Marriage and Motherhood

While marriage was consistently viewed as the normal condition of the American adult, the magazines offer abundant evidence that marriage was at the same time a Problem—a relationship fraught with difficulties that it was primarily the task of the woman to negotiate and overcome. Long before the divorce rate skyrocketed in the 1970s, marriages were perceived as being in constant danger of dissolving, and it is a rare issue of a magazine that does not offer advice on how to patch up, rejuvenate, or rescue the marital relationship. Two of the most popular series in the *Ladies' Home Journal* were "Making Marriage Work" (1947–63) and "Can This Marriage Be Saved?" (which began in the early 1940s and is still running). In 1937, fiction writer Mary Roberts Rinehart defended women in the pages of the *Journal* from the charge that marital failure was their fault, but hers was a minority view in the decades that followed. By 1940, *Good Housekeeping* was conducting a School for Brides, and a report of it in the February issue included advice on grooming and behavior so that the young wife could "play [her] role beautifully." The magazines offered quizzes women could take to predict their success as wives, and a 1948 *Woman's Home Companion* article even offered a checklist for parents to determine whether their children would be well suited for marriage.

The level of anxiety that this plethora of articles suggests was reinforced by a number of articles about divorce. A survey of a sample of *Woman's Home Companion* readers in 1944 revealed that 87 percent favored a uniform national divorce law rather than variations by state, but only 11 percent felt that divorces ought to be easier to get. This latter sentiment was echoed in articles encouraging people (chiefly women, of course) to develop realistic expectations of marriage and to work hard on their relationships instead of running off to Reno for a quick divorce. Although sectarian religion normally played little part in such discussions, the antidivorce stance of the Catholic church is thinly veiled in Reverend (later Bishop) Fulton J. Sheen's February 1953 article in *Good Housekeeping*, "How to Stay Married Though Unhappy." Sheen's message

Figure 3. Listerine advertisement from *Good Housekeeping,* August 1949.

directly counters the promise of ideality conveyed by the popular media of the era: Human relationships are necessarily imperfect; only God offers perfect love and happiness. Somewhat tangential to the concern about marriage, but clearly a by-product of it, is the sort of article that takes a stance in the debate about whether women or men had the better (happier, less stressful, more rewarding) lives. While such a debate was necessarily inconclusive, its existence pointed to both the deep sex-role divisions of the period and women's need to be reassured that their role had rewards and value.

Part of that role was motherhood, which was fraught with its own dangers. As house plans with bedrooms labeled "boy" and "girl" suggest, the ideal American family was not complete without children. The post–World War II "baby boom" may have been caused in large part by a desire for the safe "normalcy" of the family unit and by postwar prosperity, but there were other reasons as well. One of these, articulated by Dorothy Thompson in her first column in *Ladies' Home Journal* in May 1937 (shortly after a federal appeals court had overturned the 1873 Comstock Law prohibiting the dissemination of birth control information and devices), was fear of a declining birthrate among the most highly educated and, in the view of many "experts," therefore the most "intelligent" segment of the population. Thompson, apparently not cheered by the postwar increase in the birthrate, returned to this theme again in her 1949 column "Race Suicide of the Intelligent," in which she noted with dismay that young women were graduating from college to pursue careers rather than to bear children. Because statistics do not bear out Thompson's fears (nor her contention that "every field of life and activity . . . is open to women today"), it seems likely that the real impetus for such exhortations to be fruitful and multiply was the cold war: a felt need to establish a solid democratic bulwark against the threat of communism. In a 1947 article, "Let's Be Realistic about Divorce," a professor of child study articulated just this sentiment when she stated, "We know that the welfare of our nation depends on the integrity and strength of our homes."

For women who did have children, the magazines offered abundant advice on how to raise them, beginning long before a regular column by Dr. Benjamin Spock began in the *Journal* in 1954. And given the enormous dangers of improper parenting that the magazines warned of, women clearly needed all the advice they could get. Insufficient parental guidance created juvenile delinquents, while too much love and attention caused the condition that Philip Wylie, in his 1942 book *A Generation of Vipers*, termed "Momism": smothering attention that resulted in weak, dependent boys and young men. Nor was Wylie alone. In the mid-1940s a psychiatrist, Edward A. Strecker, published *What's Wrong with American Mothers?*, which was condensed for publication in the *Saturday Evening Post* in October 1946; the thesis of the book occasioned the article "Are American Moms a Menace?" in the November 1945 *Journal*. It is noteworthy that these fears about women's parenting skills center on the upbringing of male children only; there was little concern about whether little girls became sufficiently independent, and virtually the only kind of "delinquency" that seemed possible for them was the loss of virginity—although this was a dire enough circumstance to prompt Phyllis McGinley to declare in a 1954 *Good Housekeeping* article that "Unchastity Is a Sin." The fact that "parenting" was generally equated with "motherhood" is reflected in the tone and titles of the few articles that addressed fatherhood, such as "Fathers Are Parents Too" in the June 1953 *Woman's Home Companion*. But lest this trend go too far, psychologist Bruno Bettelheim cautioned three years later in *Parents* magazine that "fathers shouldn't try to be mothers" because "the male physiology and that part of his psychology based on it are not geared to infant care."

16

HELEN WELSHIMER

"My Husband Says—"

Good Housekeeping, February 1940

The favorite words of the twenty-five girls who gather around little tables each month for the sessions of Good Housekeeping's School for Brides are: "My husband says—"

For each of them has a new husband and the wish to hold him "till death do us part." And each is trying in dead earnest to learn all she can to make her marriage successful.

At the third meeting there was an especially animated exchange of what He says and what He thinks, for the brides had met to hear Ruth Murrin, Director of Good Housekeeping Beauty Clinic, talk about personal appearance and its importance to a happy marriage.

Helen Welshimer, "My Husband Says—," *Good Housekeeping*, February 1940, 45.

Here, boiled down to its essence, is what she said:

... Never become the kind of wife who says: "I don't care how I look. I'm married." One of the shocks that comes to every man who marries is the discovery that the girl he takes as a life partner doesn't always look like the girl he courted. He is bound to see you with your hair tousled, your face shining, and a streak of dust across your nose. But that doesn't mean you can let yourself go.

... Don't be the kind of girl who spends a couple of hours getting ready for a bridge party with other women, and doesn't care how she looks when there is no one around but her husband.

... Remember that the picture of you he carries in his mind all day is not how you looked last Friday night at the Country Club, but how you looked this morning at breakfast. Be sure that last glimpse of you is attractive.

... Your hair doesn't need to be elaborately dressed, but it must be smoothly brushed. You don't have to be made up, but you must look fresh and clean, awake and cheerful. Send him off with pride in his heart and a feeling that he would rather work for you than for any other woman in the world.

... Take the trouble to spruce up before he comes home. He has been looking forward to this moment all day. Make it worth while. If you drift into the habit of looking a bit messy, you can't blame him if his belief that he has captured a prize suffers a slump.

... Even if you are in the midst of getting dinner, you should be trim and fresh, clean and kissable. You don't have to be beautiful, but you should look good enough to eat.

... Be terribly fussy about grooming. Desirable men are always fiends about their own grooming, and how can you expect your husband to keep on thinking of you as a glamour girl when he knows you sometimes wear a dingy girdle?

... Don't allow your beauty methods to become an issue between you and your man. If he doesn't like to see cream on your face, use it during the day when he isn't around. If he objects to curlers, tie a fetching scarf around your head to cover them up.

... Remember that your husband has a sense of smell. Make sure that he likes the perfume you spray on so liberally. And do be careful to shampoo often enough to avoid the acrid odor of unwashed hair.

... Right now you are the pride of your husband's heart. Cherish that pride. One way to do it is to keep up the little courtesies to each other. Be at least as polite to the man you have wed as you are to an old beau whom you discarded. You married your husband because he was to you the most important and the nicest man in the world. Treat him that way.

. . . Never, never bawl him out in public, especially in front of your relatives. You think you are just making him wince a little. Actually you are deflating his confidence—a very bad thing for a man who has to fight the economic battle for a family. Too, you are showing yourself up in a bad light, and your husband knows it. You have made him feel small, but also you have made yourself look less desirable as a wife.

. . . Finally, don't grow smug about your looks. Husbands have a maddening way of saying, "I like you the way you are." They say they don't want you to diet; they don't want you to change your hair. They think your lipstick is all right. They say: "Why do you fuss with this or that?" But don't take them too seriously.

. . . If you are to keep monotony out of your marriage, let your husband look at you with a fresh eye now and then. What he really wants in his heart is for you to continue to be the leading lady of his life. And it is up to you to play your role beautifully.

17

WAINWRIGHT EVANS

"Are Good Mothers 'Unfaithful' Wives?"

Better Homes and Gardens, July 1941

A woman I know has just put her two children in a nursery school for afternoons. Ever since her babies were born—and even before—she built her life around them. She was, in fact, a sort of super-mother who refused to let anybody wait on them but herself. Her sudden decision to turn them over to a nursery school struck me as curious. I wanted to know why.

"My babies were becoming an obsession," she said. "They threw all else so out of focus that the world, including my husband, had become little better than a blur. I was becoming 99 and 44/100 percent mother.[1] I was out of touch with my friends. I didn't read any more. I hardly knew

---

[1]The very familiar slogan for Ivory soap was "99 and 44/100 percent pure," first used in 1879.

who was running for president. And as for Jim, I didn't have any time or energy left for him. He'd become a part of the furniture.

"One day a queer look he gave me when I made him turn off the radio in the middle of an important broadcast for fear it would wake the children set me to thinking. I realized how completely I was giving the children the right of way—regardless of his wishes or convenience.

"It dawned on me that Jim had been the Superfluous Man ever since the day he paced the floor outside the obstetrical room at the hospital while I did the hard work. For several years he had patiently waited for me to let our own life together swing back to normal, while I had let the children separate us. It was time to take hold of myself."

That incident of Jim and Ethel Fenwick opened my eyes to something. Luckily, Ethel was a wife intelligent enough to see what was wrong. But suppose it had been one of those hundreds of cases where the marriage begins to go on the rocks when babies arrive—especially the first baby.

Such cases are common. For it's one of the ironies of marriage that children, who are supposed to strengthen and stabilize the bond, and who do often strengthen it eventually, may nevertheless wreck a marriage.

It's true, of course, that the presence of a child serves to bind parents closer, if only for practical reasons—since care of the child provides the parents with a common job, a compelling need to work shoulder to shoulder, which is certainly one of the most valuable things in marriage. Nevertheless, this Molly and I and the baby setup can be powerfully destructive to happiness when it isn't mixed with brains. For the truth is that children often tend to separate their parents emotionally by the simple expedient of coming between them.

Children will naturally take all they can get of Mother; and the unconscious jealousy which often grows up between young children and their father for a lion's share of Mother's attention is an old story to psychologists.

One reason for all the trouble is a mother's limited strength. There isn't enough of her to go around. This is especially true while her energies are more or less depleted by the tremendous experience of child-bearing.

Later, as her strength comes back, she finds herself overdriven by a job so big that there's little time for thinking things out, and she gets to running in one groove. Many women take maternity in their stride; but others unconsciously swing over into the super-mother class. All need to guard against the tendency.

There's another reason for the trouble. When a wife becomes a mother she finds in maternity an emotional outlet which lies outside her marriage proper. That is to say, it can exist independently of her relationship with her husband. It can even trespass on that relationship. For she now has a separate source of satisfaction. She has a child to fondle and caress and nurse.

Having a baby is as deep an experience for her as marriage and the honeymoon were for both her and her husband; but this time her husband is partly excluded. What he feels is more objective. His concern is for her. His emotions toward the baby develop after the child is born, whereas her relationship with the child was a full-fledged thing long before its birth.

So satisfying does she find maternity, in fact, that it often compensates for any letdown in the romantic and intimate side of marriage.

But not so with her husband. He remains as dependent on his wife for physical and emotional satisfactions as ever; and nobody, not even that child, can take her place. She remains the controlling thing in his life simply because he's in love with her.

If his love for her is unselfish, he will realize that, so long as her strength and vitality are still at a low ebb from child-bearing, it's natural for her to give the child all she's got. The situation will usually right itself.

But if he should aggravate it by his resentment and want of understanding, then there may take place in her mind a psychological shift. She may develop a "fixation" on her children. When that happens he may never completely get her back. A husband must look out for that.

A wife must remember that even the most stable husband is likely to consider his wife's semi-desertion in favor of a more absorbing interest as a divided loyalty, and to feel somewhat as he would if she forgot him for another man. Tho he'd be shocked, doubtless, at any suggestion that she was "unfaithful" to him, yet he feels the situation to be almost equivalent.

For in a very real sense she "philanders" — and is fully supported by society and the conventions in so doing, because she is being a Good Mother. He, on the other hand, mustn't allow *her* to get out of the focus of his devoted attention for a single moment. He would be condemned by society if he should, for example, become unfaithful to her in his turn — as a sort of retaliation. This does happen. Any woman who is too good a mother to be "faithful" to her husband in this special sense of the word would do well to keep this fact in mind.

But the job of setting things right is as much his as hers. The husband with a mature point of view will fully understand the value of patience, self-control, and a certain good-humored stability that will make him a rock to lean on. His wife needs to be helped back to normal.

One mistake a great many women make is that of remaining blind to the awkward status in which every new-made father finds himself. At best he cuts a pathetic, guilty, and slightly ridiculous figure at the time his child is born; he is at the same time helpless, useless, and unhappily in the way. To put it mildly, he isn't being as important in a helpful way as he'd like to be. Tho the event has been food for humorists thru the ages, it's something more than humor. The wise wife and mother will give him a chance to feel later on that he is important and needed. She will do it by a simple trick. She will permit him to acquire responsibility by sharing actively in the care of the baby by making himself useful. And let her not get the silly idea that she is the only one with a God-given talent for tending the baby. It's quite possible that he'll prove handier and more dextrous than she.

I know a doctor who cautions new mothers when they leave his hospital: "Now don't get the idea that your baby is so important that your whole life must revolve around him. Remember you have a husband. Remember, too, that your husband is looking forward to the chance of helping take care of the baby, and that you mustn't shut him out."

One of the wisest mothers I know doesn't stop with putting responsibility on her husband and sharing his interests. She puts plenty of responsibility, too, on the shoulders of the eldest child. This gives the youngster a chance now and then to feel that he is head of the household, and responsible in the absence of his parents for the younger children. A fine preparation for adulthood and marriage!

Says this mother, "Going into bondage for your child sounds noble, but it isn't—it's silly. Too often it not only disturbs your relations with your husband, but is a means of spoiling the child and sapping his independence."

The children of that family, by the way, are beautifully mannered, independent, and able to enjoy their parents without the clinging attitude common in homes where attachment to the mother is unduly close.

One woman I know goes on trips with her husband whenever he asks her to, even tho there are three children at home, one of them still very young. Going here and there with her husband isn't a mere matter of gadding about or of social doings. It includes, rather, going on either business or pleasure trips. Such a wife falls into her husband's mental stride, and hangs onto the broad outside interests which were hers before her marriage.

She explains it this way: "I am looking forward to the future. In ten or fifteen years I won't have my children, but I shall have my husband. I'm going to see that I do have him, too. Besides, my children will be much better prepared for the time when they must cut loose from home."

A smart woman she—and a most faithful wife and mother!

18

"What's on Your *Mind?"*

Redbook, July 1945

Many worth-while short contributions come to Redbook *each month— understanding expressions of idea and opinion about the world we live in and help to shape. We welcome them, and regret that space permits us to publish so few.*

For those we find it possible to accept, we pay $100 each. . . .

People as They Really Are

So much has been written about the serious sociological problems of wartime living that a lone dissenter who says that she thinks the American people are wonderful, and who doesn't know any juvenile delinquents or unfaithful war wives, will probably not be permitted a voice.

However, I am a public-school teacher and it seems my observations might be worth something. Several of my students have both parents working in war plants. Several have fathers in service and a so-called "broken" home with a working mother. In each case, and I know the facts, the girl student goes straight home from school to do the housework her mother is unable to do, or else she has a part-time job of her own after school. The boys have after-school jobs or else spend their time on the school athletic field waiting until it's supper-time and some of their family will be home.

Most of my intimate friends have husbands in service. Most of them have at least one child. One girl, whose husband is now in France, has two little girls who always look like little dolls; she does all the fancy washing and ironing—at night, when she gets home from her war job. Her home is always immaculate. I know, because I've often spent an enjoyable evening there.

The service wives I know are young and attractive. Some have had staggering problems of fatigue, loneliness and debt to face, yet all have remained cheerful and loyal.

Don't any of the feature-writers know some of these nice average people? All we ever hear about is the other kind. It's high time for the reading public to know that teen-age rascals and unfaithful women are the exception, and not the rule!

New York

A Sure Way to Please All Women

Almost every magazine I pick up has at least one article or paragraph concerning how to hold a man, how to attract a man, how to please a man, how to keep up his morale, and so on. We women are told to let our men know how wonderful they are; we are urged to keep them aware of their superior strength, and to remind them frequently of their intellectual powers.

When a man comes home from work he should be met by a carefully-groomed, happy little woman who is all ears to learn about her husband's business triumphs. Household problems are never mentioned, of course; they are strictly taboo! After eight or ten hours of work a man must have peace and quiet and some understanding soul to hear his troubles. Also, the advisers tell us, children must be kept quiet when Daddy is around. (Are they kidding?)

Good heavens! Where is a woman supposed to get all the strength and vitality to be a leaning-post for the stronger sex? To whom does *she* go to boost her morale, to tell her troubles? After eight to fourteen hours of housework and child-care a day, where can she turn for a sympathetic ear?

Please won't some kind-hearted man write some articles on how to hold a woman, how to please her and build up her morale? Won't someone inform the men that women are human and that they need a little of the same treatment?

If some man has the courage to offer such advice, I shall be among the first to tell him how wonderful he is!

California

19

AMRAM SCHEINFELD[1]

"Are American Moms a Menace?"

Ladies' Home Journal, November 1945

"Mom" is sweet, doting and self-sacrificing. But she is not mothering her son—she is smothering him.

American mothers are so used to getting bouquets that it may come as a shock to hear that "mom" is often a dangerous influence on her sons and a threat to our national existence. "MOM DENOUNCED AS PERIL TO NATION" was the way the *New York Times* headlined it.

In brief, Prof. Edward A. Strecker, University of Pennsylvania psychiatrist and consultant to the Army and Navy surgeons general, has made this accusation: that the blame for many psychoneurotics in the armed forces, and for many neurotic rejectees, rests on their "moms," who either by overattention or stern domination during the formative years kept their sons from maturing emotionally. "Mother's boys" psychoneurotics are not to be confused with men who cracked up only under terrific battle pressures or after serious injuries, he adds. Nor are all American mothers to blame, for the majority of them are sensible, well-balanced and understanding.

Doctor Strecker is indicting the type of mother generally known as "mom." Usually she is sweet, doting and self-sacrificing, or she may be just the opposite—stern, capable and domineering. "Both these moms are busily engaged finding in their children ego satisfactions for life's thwartings and frustrations," believes Doctor Strecker. "The community applauds and fondly smiles on them. They are accorded praise and adulation for giving their lives to their children. Hidden from view is the hard

---

[1]Amram Scheinfeld (1892–1979) was a journalist, cartoonist, illustrator, novelist, and author of books on genetics for a general readership, including *You and Heredity* (New York: Frederick A. Stokes, 1939), *The Human Heredity Handbook* (Philadelphia: Lippincott, 1956), and *Women and Men* (New York: Harcourt, Brace, 1944). From 1951 to 1965 he wrote the column "Looking into People" for *Cosmopolitan* magazine, and he also published articles in *Collier's, Reader's Digest, Saturday Evening Post, Esquire,* and *Ladies' Home Journal.*

Amram Scheinfeld, "Are American Moms a Menace?" *Ladies' Home Journal,* November 1945, 36.

and tragic fact that . . . they exact in payment the emotional lives of their children."

You may recall Sidney Howard's famous play, *The Silver Cord*, in which a domineering mother refuses to accept the fact that the umbilical cord was severed at her son's birth. Remember the lines, hurled by her daughter-in-law, *"You son-devouring tigress! You and your kind beat any cannibals I've ever heard of. . . . And what makes you doubly dangerous is that people admire you—you professional mother!"*

If most American mothers were this domineering, says Doctor Strecker, so many young men would have been ruined for military service that instead of being victorious in the war "we might now be facing the prospect of defeat."

All of this involves a pretty serious charge, and must make every mother wonder about her influence on her sons. However, it is not fair to place sole responsibility for the development of a boy's character on his mother. A lot depends on the boy himself, the presence of other children (a brother or two can help a lot!) and, above all, the character and role of the father. In the majority of American homes there is a balanced situation, with a sensible mother in her proper place. But danger looms when the father's influence is weakened or missing, and the mother becomes a "mom," either of the overbearing type, who squashes her son, or the indulgent kind, who pampers and spoils him with saccharine over-attention.

This is nowhere better illustrated than in Dr. David M. Levy's recent study of "maternal overprotection." We read case histories of boys who were breast-fed up to two years and bottle-fed up to four years; who were helped to bathe and dress up to the age of fourteen; who were led by hand to school and kept from other boys so that they "wouldn't learn bad things" or "play rough." And there was frequent kissing and fondling and, in some cases, sharing of the mother's bed up to the boy's adolescence. Some of these boys were turned into helpless ninnies, with no will of their own, while others became young tyrants.

Significantly, in half of Doctor Levy's case histories the boy is an only child. Doctor Levy also notes that in many of the "overprotected" situations the father is apt to be weak-willed or submissive (often because he is a "mother's boy" himself) and leaves the handling of the child to his wife. But in the majority of cases, the "problem child" has a "problem mother." Frequently her oversolicitousness and undue attachment to her sons result from her inadequate sex life, which may cause her to seek in a son a love substitute. Again, if her own ambitions have been suppressed, she may strive to achieve success through her son, driving him beyond his powers.

The "mother's boy" reveals himself by these traits: He is apt to be hypersensitive, worrisome, fussy, vain and somewhat of an exhibitionist. He basks in attention and sulks if he doesn't get it. Although he strongly desires affection and is very eager to make friends, he may not hold on to them. Being sensitive and intuitive to a high degree, he may be very understanding, sympathetic and anxious to help others, which often leads him into public life, charitable work or social service. On the other hand, if he is ambitious and calculating, the same faculty of sensing people enables him to exploit them to his personal advantage.

The victim of maternal overprotection may be impelled toward great heights of achievement or toward depths of failure, depending upon the mother's character and the boy's inherent capabilities. History is full of notable "mother's boys." We could start with our late President Franklin D. Roosevelt, whose lifelong attachment to his mother and her profound influence on him are well known. He was an only son, whose father died when he was eighteen, and up until her death a few years before his, his mother was one of his closest confidants. Another great president, Abraham Lincoln, said, "All that I am or hope to be, I owe to my angel mother." President Harry Truman also appears to have been greatly influenced by his mother. She was an active campaigner for her son when he first ran for the Senate, and at ninety-two is still much in the foreground. "The resemblance between mother and son is deeper than a physical likeness," wrote one reporter recently.

In the field of literature, a few pronounced mother's sons who come to mind are Byron, Goethe, Victor Hugo, Gustave Flaubert, Oscar Wilde and, more recently, Thomas Wolfe and George Bernard Shaw. D. H. Lawrence's novel, *Sons and Lovers,* was based upon his own tragic experiences. The theatrical profession abounds in examples. I can particularly remember David Belasco, the producer, as he chatted with me in his exotic studio, attired in clerical garb, his every inflection, gesture and movement the consummate revelation of a mother's darling.

It is worth noting that although some mom-coddled boys achieve great success, many are nonetheless socially maladjusted and personally unhappy. Stephen Foster, America's great song writer, was a much-pampered youngest child who never outgrew his emotional dependence on his mother. He finally became an acute alcoholic, deserted his wife and child, and died a broken Bowery derelict.[2]

---

[2]The Bowery is a street in Lower Manhattan noted at the time for its saloons and squalid living conditions and populated by derelicts. The word has come to be used for any area with similar conditions.

Among abject failures, we find a high proportion of mother's sons. When a man fails to win from society the plaudits his mother has led him to expect, he may feel that the world is discriminating against him. So it is that mother's boys abound among the social misfits, ne'er-do-wells, criminals, alcoholics and homosexuals. Adolf Hitler was the only son and spoiled darling of his not-too-bright mother, whom some biographers report as "the only woman he ever loved." She alone indulged his ambition to be an artist, at which he failed miserably; and, as in the case of many another mother's boy, the fact that Adolf expected continued coddling and admiration from the world, and was cold-shouldered instead, may explain his urge to destroy the existing scheme of things and make it over to suit himself. Had this one individual had a different mother, history might have taken another course.

In their relationships with women, many mother's boys continue the trend toward extremes. If successful in amours, they may become Don Juans and Lotharios, going from one romance to another in search of emotional satisfaction which the ever-present mother fixation keeps them from attaining. To quote Dr. Otto Fenichel, "The Don Juan seeks his mother in all women and cannot find her." However, if the mother's boy is not appealing to women or fears he won't be, or has an "incest" feeling regarding opposite-sex relations because of identification with his mother, he may shun women altogether and frequently turns to homosexuality.

Reporting on the background of many male homosexuals, Professors Lewis M. Terman and Catharine C. Miles say in their study, *Sex and Personality,* "The psychosocial formula for developing homosexuality in boys would seem to run somewhat as follows: too demonstrative affection from an excessively emotional mother, especially in the case of a first, last or only child; a father who is unsympathetic, autocratic, brutal, much away from home, or deceased; treatment of the child as a girl, coupled with lack of encouragement or opportunity to associate with boys and to take part in the rougher masculine activities; overemphasis of neatness, niceness and spirituality; lack of vigilance against the danger of seduction by older homosexual males."

The formula need not always work, of course, for even from the environment described many boys grow up to make normal sexual adjustments. But in any event, the mother's boy may remain a bachelor for a long time or permanently. An Army psychologist told me, "We're afraid that a lot of unmarried G.I. Joe's who put their moms on pedestals may have developed mother fixations while overseas, and, now that they are coming back, they may have difficulty in finding wives who will measure up to their overidealized notions of womanhood."

In recent years, many factors have contributed to the overbalance of maternal influence in the American home. Where fathers have been away in service for prolonged periods, little boys are started off with a one-sided mother attachment, and if the father fails to return, the mother may seek to make up for the loss through her sons' affections. Apart from this, the trend toward more mother's boys has gained impetus from the decrease in family size, resulting in more only sons. Moreover, various changes in American life are tending to keep more fathers away from their homes and to leave young sons largely to the mothers, with nurses, domestics and women schoolteachers adding to the feminine atmosphere. Another factor has been the ever-higher survival rate of wives compared with husbands. This has led to an increasing number of widows, and hence to more control by mothers over the family purse strings and business interests, increasing their sons' dependence on them economically as well as emotionally.

The growing one-sided influence of American mothers is more than a matter of individual concern. It involves significant changes in our whole national culture and outlook, for, as authorities have pointed out, the thinking, the behavior and the achievements of nations can be molded by the relative degree of mother or father influence. As examples, the Irish, and other Catholic peoples generally, are strongly influenced by mothers, chiefly because of reverence for the Virgin Mary. The English and other Protestant nations tend largely toward father domination. The Germans in recent generations went to extremes in heightening the father influence and repudiating mother influence, and the viciousness and brutality of the Nazis reflected this. The Japs, too, have exalted the father influence and suppressed that of the mothers. Conversely, Chinese women have continued to be strongly influential, as revealed by the roles of Mme. Sun Yat-sen[3] and her sisters. (Prof. Ray E. Baber reports with respect to the Chinese family, "In any dispute between a wife and his mother, a man was supposed to take the mother's side, on the grounds that he could get another wife but could never get another mother.") We might also note what happened when the Italians, a strongly mother-influenced people, tried to emulate—somewhat pathetically, it is true—their warlike forebears, the Romans, a father-dominated people.

So it can be seen that unbalanced father influence, either in the individual family or the nation, can be as bad as too dominant mother influence. In this country, however, the growing domination by American mothers of the "mom" type is the more immediate menace to our secu-

---

[3]Sun Yat-sen (1867–1925), early leader of the Chinese Nationalist Party. His second wife, Soong Ch'ing-ling (1890–1981), became an important figure in the People's Republic of China and a leading proponent of communism.

rity. Looking backward, one could readily grant that a good deal of the strength of America came from the mothers who worked side by side with their menfolk, created and managed homes under innumerable hardships, gave of themselves unselfishly and reared their sons to assume the responsibilities of men. It is on these women that the tradition of reverence for the American mother has been built. No one would deny that a majority of American mothers today, adapting to the needs of changing times, are deserving of similar respect and praise. But if we are to believe the psychologists, there is a very large number of mothers who have not earned the right to wear the laurels won by others, or to accept smugly homage for service they have not rendered.

If we have indeed gone too far in the direction of maternal conditioning, we must see that the psychological diet of our boys is supplemented by more "masculine vitamins." American fathers must be impressed with the need of greater participation in the rearing of their sons. Every mother can agitate for more male teachers in our elementary and high schools, and she can encourage manly activity for her boy through the Boy Scouts, other boys' groups and young men's organizations. Compulsory military service at the age of eighteen would also help, some authorities feel.

Strangely enough, it is in the more privileged and enlightened groups that the situation needs most attention, for it is among these that only sons are most common, that psychological tension is apt to be greatest, that fathers are the most neglectful and mothers most inclined to be domineering. Prof. Arthur T. Jersild, child psychologist of Columbia University, told me, "Where the more ordinary women take the rearing of their sons in stride, many so-called 'sophisticated' mothers are so eager to do a thorough job of mothering that they wear themselves and their boys to frazzles." Or, as a highly neurotic man patient told one of my psychoanalyst friends, "My trouble is my mother didn't mother me—she *smothered* me."

The best way to avoid becoming a "mom" or having a maladjusted son is to follow the ten don'ts that are listed below. And take it easy. Boys will be boys—if you just let them be.

Don'ts for Doting Mothers

(IF YOU WANT YOUR BOY TO DEVELOP NORMALLY)[4]

1. Don't breast-feed or bottle-feed your boy any longer than absolutely necessary, and don't dress or bathe him beyond the time that he can care for himself.

[4]Compiled from suggestions by various authorities. [Note in original—ED.]

2. Don't have him share your bed after he outgrows babyhood.
3. Don't treat your son like a lover. Avoid excessive fondling and kissing (particularly "mouth" kissing).
4. Don't get your son in the habit of letting you make his decisions for him.
5. Don't rear him in an exclusively female atmosphere (if there's no father on the scene), but see that he has plenty of opportunity to be with adult males as well as boys his age.
6. Don't force your son beyond his capacities, or try to make his success the compensation for your own failures.
7. Don't whine and complain (as he grows older) that he's neglecting you, or doesn't love you enough.
8. Don't make him feel you are jealous of his girl friends or that they're competitors of yours.
9. If you have a son and daughter, don't show him favoritism at her expense.
10. If you are widowed, or divorced, don't try to turn your son into a substitution for your husband, or make him feel that he will be an ingrate if he marries and has a home of his own.

20

"Are You Too Educated to Be a Mother?"

Ladies' Home Journal, June 1946

The Census Bureau recently completed a five-year survey and published these findings: A woman who has gone to college is likely — on a national average — to have a fraction more than one child; the likelihood that she will have even one child is growing slimmer. If she is a high-school graduate, chances are she will have two children. But the woman who left school in the fourth grade is almost certain to have at least four.

This difference in birth rates is probably a direct result of use or nonuse of birth control. In one representative community it was found that 80 per cent of the "well-to-do" women practiced birth control. Only 30 per cent among the "very poor" used contraceptives. Birth control,

"Are You Too Educated to Be a Mother?" *Ladies' Home Journal,* June 1946, 6.

however, is merely a *means* of expressing the desire for a small or limited family. Desire for fewer children can be traced to the high cost of education, insistence on a high standard of material living and competing interests outside the home.

The make-up of our future population is frightening to some students of the rise and fall of civilizations. At the present rates of reproduction, within three generations the woman with a bachelor degree will have *one* grandchild. Her contemporary with less than a grammar-school diploma will have *nine* grandchildren. Geneticists state flatly that *"there is danger of outright decline in the physical and mental make-up of our population. We are not reproducing the best of ourselves."*

In the geneticist's dictionary, "the best" is not a term of social snobbery, but a matter of scientific evidence—evidence that children of college-educated parents have a higher *average* intelligence than children born into non-educated homes. Thus, our educated women, potentially mothers of children with greater native ability, are guilty of squandering their genetic inheritance. Unthinkingly they are lowering the standards of future generations.

The first ominous signs are now appearing in our public schools. In one large Eastern city, the standard I.Q. test has been arbitrarily scaled down to meet the declining quality of the students. A child who would have rated only 90 on the test ten or fifteen years ago is now graded at 100. Thus the schools are hiding unpleasant facts from wide circulation.

There may be a negative comfort in knowing that other important nations are facing the same problem. There is positive encouragement in Sweden, where the unfavorable birth ratio has been *reversed.* Women in Sweden's upper education and income brackets are now scoring the highest birth rates, while women of less education and financial resources are bearing fewer children. Sweden produced this turnabout by a two-fold program. The cost of bearing and rearing children has been reduced, through free obstetrical services, nursery schools, dental clinics for school children. For those who *do not* want children *at any cost,* birth control is available on a nation-wide basis.

Sweden wisely realized that state aids alone—no matter how generous—would not bring about any great increase in births unless there was a psychological upgrading of the value of children in family life. In our own country this change of mind would be a major factor in adjusting the downward population spiral—if we should decide to do anything about the situation. Even as a nation of avowed individualists we will have to take the future into account as well as the whim of the moment. We must learn to take our babies more seriously and less sentimentally. We must learn that our educational opportunities are not an outright gift, to be prodigally misspent. We must learn that we are not too educated to be parents; we must learn that we are too educated *not* to be!

21

"What Makes Wives Dissatisfied?"

Woman's Home Companion, April 1947

Every wife is bothered by certain aspects of her marriage. That is only human. She still can be gloriously happy. Even as I was preparing this article, a young wife I know gave me evidence of it. Looking over my shoulder she emphatically assured me that she could fill a book with causes for dissatisfaction. "I get so mad when he rinses his hands and then wipes them clean on the towel," she said. "And what can you do with a man who persists in strewing his business papers all over the piano?"

Actually she is happily married. She was giving vent to momentary irritations, not to basic dissatisfactions that undermine a marriage. She did not distinguish irritants from deep-seated maladies.

But how can a woman tell the difference with certainty? What are the conditions that make a wife become basically dissatisfied?

To find out, I interviewed one hundred truly dissatisfied wives. By their own admissions and by my own test scores they ranged from "somewhat" to "decidedly" or "extremely" unhappy.

Who are these one hundred wives? Their average age is twenty-five. Fifty-six are married to war veterans. They have been married an average of two and a half years and have been unhappy, on an average, since the eleventh month of their marriage.

The explanations they gave for their unhappiness are not what you'd expect. Not one complained of her husband's untruthfulness, tightness with money, drinking or profanity. And just one complained of her husband's slow progress in his career.

More startling, only ten suspected their husbands of infidelity and of those only six gave it as a cause of their unhappiness.

Furthermore, their explanations were strikingly different from those given by four hundred husbands I interviewed earlier. The husbands' dissatisfaction with marriage centered around their wives' lack of emotional control. They listed such faults as nagging, jealousy and extravagance. The complaints of the dissatisfied wives were much more fundamental.

The wives were distressed because there was little sharing, little comradeship, little of the spirit of partnership they had assumed would come with marriage.

Women in general are more likely to worry about their marriages—and become dissatisfied with them—than men. The fact shows up in all surveys. This is so partly because women are more introspective than men, partly because marriage looms larger in their lives. The woman who becomes dissatisfied with her marriage becomes dissatisfied with her whole life. The husband who becomes dissatisfied can sometimes shrug it off.

But even allowing for the general tendency of wives to worry more than husbands, the one hundred unhappy wives I interviewed gave convincing reasons for their dissatisfaction. All centered around one unmistakable condition—the failure of the husband and wife to achieve a genuine partnership—but were in six broad categories:

1. *The lack of a confidential relationship.* Perhaps the greatest value of marriage is the opportunity it gives husband and wife to share their troubles, ambitions and triumphs. Yet sixty-five of the one hundred wives confessed they did not feel free to talk over "everything" with their husbands. Twenty-eight specifically blamed the fact for much of their marital trouble.

Many happy wives also lament that their husbands crawl into a shell when they come home from the office. I remember one charming and happy wife who revealed her chief complaint with marriage when she said, her eyes snapping: "He won't talk when I want to talk!"

One third of the unhappy wives, however, said that their husbands did more than crawl into a shell—they were actively annoyed by an attempt to discuss the problems of the day. A newlywed said, "He just gets up and walks out in the middle of my sentence."

Because the confidential relationship is lacking, most of these wives and husbands have never worked out an easy system for settling differences. They don't know how to compromise, which is a prime requisite for any happy marriage. Every difference means a quarrel, and eighteen of the wives say they "always have to give in" to restore peace.

"My husband always threatens to leave," one explained, "if he can't have his way. He has a superiority complex so far as I am concerned and lords it over me."

But another wife showed her own lack of insight when she wrote on her test: "He will never admit mistakes even when I prove him to be wrong." Virtually no human being, male or female, can really accept the fact he is clearly in the wrong. He must be left some justification. The smart—and happy—wife knows how to get her way without arguing or trying to show her husband how wrong he is.

2. *Little meeting of minds.* To really enjoy each other's company it is important for husband and wife to be compatible on the intellectual level. They should have many of the same serious interests and at least roughly similar philosophies, religious views and formal education.

Most of the hundred unhappy wives stated flatly that such compatibility did not exist in their marriages. And they blamed its absence for much of their dissatisfaction.

In answer to the question "Are you and your mate intellectually well matched?" thirty-nine said no, perhaps in part because only half of the wives had attended college while virtually all the husbands had.

Thirty said they and their husbands disagreed either "frequently" or "always" in their philosophies of life. One complained of her husband's "different moral code." Another described her husband as "cynical and blasé," herself as "idealistic."

The difference in philosophy showed up not only in conflicting outlooks on life but also in the wives' disapproval of certain of their husbands' habits such as poker playing, betting on races and having friends with low standards.

And when I asked the hundred wives what qualities they would particularly like to change in their husbands, more (forty-three) checked religion than any other quality, although fewer than a dozen are married to men of radically different religious groups. The trouble seems to be that the husbands are agnostic or indifferent to religion while the wives place a great deal of faith in spiritual values. One wife said: "I think we should make some effort to get to church oftener but he finds all sorts of excuses."

3. *Failure to enjoy life together.* The picture of a man and woman walking hand-in-hand together down the avenue of life is a familiar symbol of marriage. But such a stroll is unfamiliar to most of the hundred unhappy wives.

Thirty-seven explicitly expressed their chagrin at finding out after marriage that they and their husbands shared few interests. Many others complained that although interests were shared everything was spoiled by bickering about the details—where to go, how much to spend and when to come home. Twenty said they and their husbands could rarely agree on recreation.

Apparently much of the inability of the hundred wives to enjoy life with their husbands springs from their basically conflicting personality patterns. I mean particularly the marriage of an extrovert and an introvert, always an unfortunate combination. In fifty-two of the hundred marriages wives who "like to be on the go" were married to "stay-at-home" males, or vice versa.

4. *Cleavages in relation to outsiders.* By outsiders I mean primarily in-laws and friends. Perhaps the housing shortage has forced too many young couples to share a roof with in-laws because twenty-nine of the hundred wives complained that in-law problems were doing much to make their marriages unhappy. That to me is a surprisingly high number. The troubles seem to center mainly around lack of privacy, sharing expenses and unwanted suggestions.

If a couple must live with in-laws I feel it imperative that they pay a definite rent so they won't feel obligated. Still more important, neither partner should discuss his marital tribulations with the in-laws.

Fourteen wives reported disagreements with their husbands about friends. Most of the resentments here arise from the fact that the wife disapproves of her husband's behavior while with these friends. He stays out too late, meets flirtatious women, lends money that isn't paid back or neglects his wife in favor of friends.

5. *Conflicts in running the home.* A large proportion of the hundred wives report they are dissatisfied with their husbands' attitude toward the home.

Thirty-six said they and their husbands cannot agree on the management of the family money. A very unhappy wife said her husband was forever embarking on "financially expensive brainstorms without consulting me."

Forty said they have occasional conflicts with their husbands over the rearing and discipline of the children. One aggrieved mother said: "He takes no responsibility for our baby at all."

Fifteen said their husbands are simply not interested in the home.

6. *Failure to find physical enrichment.* When husband and wife are unable to cooperate harmoniously in day-to-day living it is understandably hard for them to pretend to be lovers. There can be no genuine affection for a person you are constantly annoyed with. But most of the hundred unhappy wives were still trying to pretend. Although eight had stopped kissing their husbands altogether, seventy-one kissed theirs each day. Several mentioned, however, that the kisses had become casual and somewhat guilty pecks. A wife who had been married two years said: "I have no more romantic illusions. We've just about forgotten what it is like to kiss each other and really mean it."

Thirty-nine of the hundred expressed a deep yearning for more affection. They kiss and engage in full intimacy with their husbands, but they sense an absence of spontaneity and yearn for verbal assurance that they are loved. To a woman a caress often has little meaning unless accompanied by genuinely tender words. Careless husbands overlook this fact.

It is significant that to the question "How well adjusted sexually to each other are you and your husband?" only fifteen wives answered favorably without reservations. Fifty-five responded that they were definitely *not* well adjusted.

Here are some of the other facts we learned about the "love" lives of the hundred:

Forty-eight have been unable to agree with their husbands about the frequency of intimacy.

Thirty do not find their husbands attractive sexually, or only mildly so.

Nine have never experienced full satisfaction in intimacy. Thirty others have experienced it only rarely or occasionally.

The basic grievance was voiced by a wife who wrote: "My husband has never made any effort to understand my feelings about the matter. Frankly I have not found it a sublime experience."

Summary. When all the dissatisfactions, annoyances and complaints of the hundred wives are analyzed it is evident that many of their maladjustments are merely symptoms of basic troubles. Again and again the evidence suggests that they are unhappy primarily because of:

— Unlike dispositions, attitudes and personality traits, which have produced psychological incompatibility.

— Emotional immaturity, of self, partner or both. The scores of the wives on our stability test indicate that they are, as a group, more unstable emotionally than average, and are in many cases neurotic.

Instability caused many of the wives to plunge hastily and romantically into an unsound marriage in the first place—and has prevented them from taking sensible steps to achieve a satisfying relationship in spite of their shortcomings.

Many couples I know have built enduring marriages despite personality differences, in-laws, miserable housing, financial stresses and the rest. But they adopted a mature attitude and regarded their marriage as a contract. They unreservedly accepted its duties and responsibilities.

Any wife whose marriage is not as happy as she wishes should not waste time trying to find who is to blame. Without delay she should:

First, talk out the entire situation with her husband in a calm and helpful spirit. Find where the trouble is. I call this process mutual psychotherapy. It does wonders in reducing tensions and clearing the air of misunderstandings. Alone if necessary but with your husband if possible, try to work out steps you can take to remove the irritants. A wife's most important objective should be to recapture a confidential relationship with her husband. After that everything will seem easy.

Second, resolve that above all else in your life you are going to make your marriage work. This determination in itself will enable you to overcome most obstacles. When both partners had that determination I have never known a marriage to fail.

Be sure to do *more* than your share and do not let false pride hold you back. Wives often successfully reestablish a warm relationship merely by showing less restraint in receiving and reciprocating demonstrations of affection. If you are generous and warm-hearted, your husband will respond in kind. Strive to save your marriage, not to save your face.

22

CLIFFORD R. ADAMS[1]

"Making Marriage Work"

Ladies' Home Journal, January 1948

Is it you or your marriage that gets out of hand? Check here and you'll soon find out.

Facing Your Needs

Success in making friends or holding a job largely depends on a well-adjusted personality. So does happiness in marriage. And the first step toward a well-adjusted personality is to face your needs, then make an honest effort to satisfy them.

[1]Clifford R. Adams was a professor of psychology and director of the marriage counseling service at Pennsylvania State College (now Pennsylvania State University). This article was part of a series titled "The Companion Marriage Clinic." Adams's column "Making Marriage Work" was published in *Ladies' Home Journal* from 1947 to 1962, when it was supplanted by the feature "Can This Marriage Be Saved?," which still appears in each issue.

Clifford R. Adams, "Making Marriage Work," *Ladies' Home Journal,* January 1948, 16.

Our physical needs are obvious; if we lack air, water, food or shelter, we perish. Our psychological needs are just as vital. When they are not satisfied, emotional suffering follows—even suicide.

Alice Rand is an unhappy woman, and she and her husband Bill have a thoroughly unhappy marriage. He compensates for unsatisfactory home life by putting all his energies into his job and getting ahead. Alice just stays at home, complaining that her two children keep her tied down.

Alice is not facing her needs; instead, at 31, she is filled with self-pity. Though she tells everybody she never has a spare minute, she finds plenty of time for soap operas, confession magazines and games of solitaire. And her widowed mother would gladly stay with the children so she could get out more.

She tells her husband that nobody is friendly, and that she is miserable. But has she returned calls that were made when they moved into the neighborhood? Does she go to church, or to club meetings? Has she learned to play bridge, a game Bill enjoys? She's done none of these things.

Perhaps Alice can be shown what is wrong. To attain the happiness she longs for, she must face her needs and try to satisfy them. If she can accept this idea, she will begin to enjoy life—and so will her husband.

Alice is not unique. A wife who mopes, cries or has temper tantrums has not adjusted to her environment. In order to become adjusted, she must recognize and meet her needs.

What are these major motives or needs? Here are five essentials to mental fitness:

Social approval. Every human being needs the feeling of being accepted, welcomed and respected. Without the appreciation and praise of friends, your achievements seem empty, your life is lonely. But to win the approval of others, you must reward them in kind. Unless you are friendly and appreciative yourself, your friends will be few.

Belongingness. You feel secure only when you know that you *belong* to some group, that you are an integral part of your community. To be part of your community, you must take part in its affairs. Church membership, community activities, simple neighborliness, all are ways of belonging.

Mastery is an important motive because it brings a sense of fulfillment. If this need be thwarted, the individual feels that he is falling short, and becomes surly, resentful and quarrelsome. Everyone has opportunities to satisfy this motive. Managing a home, heading a committee, directing helpers, all are acceptable ways of satisfying the need for mastery, for all are means of expressing your own will, of accepting and discharging adult responsibilities.

The need for love and affection begins, but does not end, in infancy; throughout life the adult nature continues to need the assurance of love. But love is an obligation as well as a reward. To receive love and affection, you must give them without stint.

Sexual satisfaction. Although this complex need has a physiological basis, its development is psychological. Playing with other children, dating, youthful flirtations, all are natural steps toward satisfying this need. Marriage, the ideal basis of the sex relationship, is also the avenue to a full and well-balanced life. For marriage offers the one best way of meeting most of our needs.

You are starting a new year. List these five needs, and any others of importance to you. Then list the ways you are satisfying them. If something is lacking, see what you can do to correct the situation. These hints may help you:

— Plan your life. Budget your time and make provision for ample social activities.
— When a problem arises, get the facts. See what can be done— and *do* it.
— Learn to laugh at yourself; encourage a sense of humor.
— Talk over your problems with a close friend, preferably your husband if you are married.
— If you're dissatisfied, restless, bored, don't waste time feeling sorry for yourself. Join a club, cultivate new acquaintances, develop a hobby. It's up to you to do something to change your routine.
— Mental and physical health go hand in hand, and both are priceless. Guard them closely!

Making Marriage Secure

Women, single or married, are less secure than men. Women have led more sheltered lives, have had fewer opportunities to acquire independence. Consciously or not, women seek security in marriage. A man thinks of marriage first as a partnership, and of his wife as a companion. But to a woman, a husband means a home and children and the inner security she craves.

There can be little happiness in a marriage without emotional security: it is a factor in at least four out of five unhappy marriages discussed with me. And in two thirds of these, the problem is directly related to financial difficulties.

That is why the wise married couple, seeking happiness in marriage, consider money management not merely as an end in itself, but as a step toward the satisfying relationship they desire.

It is the management of money, not the amount, that causes trouble or prevents it. How the money is spent, and for what, is far more important than how much there is, although either too much or too little can cause trouble. Once actual living expenses are assured, how can your family's income be managed to provide maximum security?

— First on the list is health insurance. Too often families make no provision for the expense of illness, and are crippled when it strikes.

— Next comes life insurance. A husband's insurance should not be less than the total required to support the family for a year. In addition, enough should be provided to cover funeral expenses for each member of the family.

— A savings account of $500 should be built up for emergency situations. Double that amount is better, with some of it in Government bonds.

— Then might follow additional home furnishings, conveniences and comforts. Before buying a car, remember that the minimum cost of depreciation, running expenses and repairs is a dollar a day—more than 10 per cent of the average family's income. Unless your income is greater than average, you cannot afford a car without sacrificing many other comforts or even eliminating savings. Better postpone it, unless a car is essential and will partly or wholly earn its own way.

— Home ownership is a common goal. Though it probably costs a little more to own than to rent, it is worth it to most families for the sense of belongingness and security that a home of your own brings. How much should be spent for a home? Never more than three times, better still only twice the average annual income. Rent (or mortgage payments) per month should not exceed a week's salary.

— Whatever plan you adopt, you and your husband should agree on it jointly. The management of money, like any other family problem, is a responsibility for husband and wife to share.

Do You Agree?

My fiancé is rapidly becoming an alcoholic. We have postponed our marriage because of his drinking. Is there any way he can be helped?

Not until he wants to be. Does he honestly want to *stop* drinking? Can he be persuaded to join Alcoholics Anonymous? Will he consult a psychiatrist? If so, there is hope. Psychotherapy and some of the recent medical approaches (insulin, for one) may help. Don't marry until a year has passed since his last drink. Marriage never reformed an alcoholic — or any other escapist.

Ask Yourself: "Is My Marriage Happy?"

Perhaps your marriage is happier than you think. Study each question before answering *Yes* or *No.*

1. Does either have habits to which the other objects?
2. Is anything in your marriage especially unsatisfactory to you?
3. Do you ever wish you had never married?
4. Would you change your mate's disposition if you could?
5. Has your husband ever talked of separation (or divorce)?
6. Is your present housing reasonably satisfactory?
7. Do you *rarely* have misunderstandings about money matters?
8. Are you two about equally loving and affectionate?
9. Is your love deeper now than when first married?
10. Are both of you free from jealousy or distrust?
11. Do you live without interference from in-laws?
12. Are you two well-adjusted sexually?
13. Do you have much the same ideas about children?
14. Are occasional quarrels made up the same day?
15. Do both enjoy much the same social activities?
16. Can you freely talk things over without reservations?
17. Do you two have many mutual friends?
18. Are you free from debt and financial stress?
19. Is love *mainly* why you live with your mate?
20. Do you believe your marriage to be successful?

The first 5 questions should be answered *no,* and the last 15, *yes.* Fifteen or more correct answers almost prove your marriage is happy. A score of 11 to 13 is average. If you scored less than 10, analyze your answers with your husband to see how conditions can be improved.

23

MRS. DALE CARNEGIE[1]

"How to Help Your Husband Get Ahead"

Coronet, January 1954

There are many simple ways in which an astute wife can give her mate a powerful push up the ladder to success.

Not long ago, an old friend dropped in to see us. He looked tired and unhappy.

"I don't know what to do," he told us. "For six months I've been working overtime trying to develop a new branch of our business. I get home late every night. Once this spadework is done, I can get back to normal hours. But Helen is so unhappy about not having me home for meals and our never going out together that it's getting me down.

"Establishing this new line is important to both of us, but I haven't been able to make her see it that way. I worry so much about her, I can hardly keep my mind on what I'm doing."

Such periods of intensive labor at some out-of-the-ordinary task are no picnics for wives, however necessary or fascinating such work may be to their husbands. We wives have to stand by as bodyguards, nurses and morale-builders—gritting our teeth silently and wondering if we will ever lead normal lives again. We have none of the thrill of achievement that motivates our mates and makes them deaf, dumb and blind to everything but the job in hand.

In this situation, if you want to help your husband get ahead, plan some diversion for yourself to keep from brooding over how different things used to be. Learn to carry your own weight socially instead of depending on your husband's presence to make you a desirable guest. There are many situations where you won't fit in as an extra woman; avoid them. You will be welcome as May sunshine in others.

---

[1]Dale Carnegie (1888–1955) was the author of *How to Win Friends and Influence People* (New York: Simon and Schuster, 1936) and other motivational texts, beginning with *The Art of Public Speaking* (New York: Association Press, 1915) and ending with *How to Stop Worrying and Start Living* (New York: Simon and Schuster, 1948). This article, by Carnegie's second wife, is from a book with the same title that she published in 1953.

Mrs. Dale Carnegie, "How to Help Your Husband Get Ahead," *Coronet*, January 1954, 65.

Try doing some of those things you never had time for before; visit some art galleries, go to a concert, do some work for your church or political party. Try a self-improvement course or some night-school classes. Such a program will do you good and keep your husband from worrying about your being lonely. Remind yourself that this is only a temporary situation. If you prove you can take it in stride, you can have a second honeymoon when the big push is over.

If you have a job or career of your own, would you be willing to give it up if it would advance your husband's interests? If not, you are more interested in promoting yourself than promoting your husband.

Helping a man attain success is a full-time career in itself. You can't hope to do it unless it is important enough to claim all your attention.

Beautiful, blonde Zetta Wells, wife of famed explorer Carveth Wells, had a fascinating career of her own when she met her husband-to-be. Zetta was a successful radio and lecture manager who looked after the business interests of many famous people. Carveth Wells came to her as a client, fell in love with her and married her—on Zetta's condition that she be allowed to keep her exciting job and her prized independence.

The marriage took place in March. In June, Carveth was leaving for a trip to Russia and Turkey to climb Mount Ararat. Zetta expected to stay home and work. But when the time came, she couldn't bring herself to remain behind. Just this once, she said. So off they sailed on an adventure that turned out to be a nightmare of hardship and frustration— although it brought forth Carveth's best-selling book, *Kapoot.*

Zetta's job, when she came back to it, looked pretty tame in comparison. So a year and a half later she was off with him to Mexico to climb Mount Popocatepetl. This, too, was a grueling physical ordeal. Zetta was cold, hungry, exhausted and scared silly most of the time. But she was thrilled, too.

The winds of that mountain peak blew away the last shreds of Zetta's die-hard independence. She realized that being Carveth Wells' wife was worth more to her than any amount of success she could win on her own.

When they returned from Mexico, Zetta closed her office. She was free to follow her husband to the ends of the earth—and that is exactly what she did.

I do not underrate the many wives and mothers who are forced by circumstances to work at jobs outside their homes. I believe that women should equip themselves to earn a living by their own efforts, since life is uncertain. But since we are discussing ways by which wives can help their husbands to succeed, we cannot ignore the fact that this is a big enough job in itself to demand single-mindedness and full-time effort of a wife.

Whatever a man's occupation, his chances of getting ahead are increased by his wife's ability to get along well with others and her skill in adapting to social demands. If this ability comes natural to her, so much the better. If not, she must acquire it.

Don't think that because your husband is now filling a somewhat lowly position, nothing is expected of you. The business, industrial and professional leaders of tomorrow are all unknown, obscure young men today. Nobody starts at the top. Are you prepared to do your husband credit 10, 20 or 30 years from today, when he is a leader?

Start today! If you have fears, prepare to shed them now. If you are awkward or tactless, learn to love, respect and enjoy other people. If you feel a lack of educational background, don't hide behind that threadbare excuse, "I never had a chance to go to college." Take courses in night school. If you can't afford that, run, don't walk, to your nearest public library.

Learning to make—and keep—friends and to get along with others is one basic way to prepare for the time when your husband achieves a position of importance. If he is clumsy in handling people, a tactful wife will help make up for his blunders; if he is diplomatic in his human relations, a wife must be, also—to keep him from looking ridiculous, if nothing else.

Wives have been trying to influence husbands by nagging since the days of the caveman. Such differing personalities as Napoleon III and Abraham Lincoln were afflicted with nagging spouses.

Women are still trying to make nagging pay off. To date, it hasn't worked—except in reverse. Dr. Lewis M. Terman, psychologist, made a detailed study of more than 792 marriages. Results showed that husbands ranked nagging as the worst fault a wife could have.

An old friend of our family told us that his career was almost wrecked by a wife who belittled every job he ever had. He started out as a salesman. He liked his product and was enthusiastic about selling it. But when he came home at night, his wife would greet him saying: "Well, how's the Boy Genius? Did you bring home any commissions or just a lecture from the sales manager? I suppose you know the rent is due next week?"

This went on for years. In spite of it, the man did forge ahead by sheer ability. Today he is an executive vice-president in a nationally known concern. His wife? Oh, he divorced her and married a younger woman who gives him all the affectionate support denied him by his first wife.

Complaining, whining, comparing, sneering, harping—the nagger may specialize in one or be a general practitioner of all these forms of mental cruelty. The bride of 20, who confines herself to a few digs about when are they going to be able to have a new house like the Martins', is, at 40, a chronic, unlovely complainer who is never satisfied with anything. Nagging is a devastating emotional disease. If you are in doubt about having it, ask your husband. If he should tell you that you are a nag, don't react by violent denial—that only proves he is right. Instead, take steps to correct the situation. Here are six suggestions that may help to cure it.

1. Enlist the cooperation of your husband and family. Ask them to fine you 25 cents every time you show irritation, give a harsh command or harp on a sore point.
2. Train yourself to say a thing once only—then forget it. If you have to remind your husband peevishly six times that he promised to mow the lawn, he probably isn't going to do it anyway, and nagging only makes him balky.
3. Try to get results by softer methods, like "If you will mow the lawn, honey, I'll bake your favorite pie for supper."
4. Cultivate a sense of humor. It will give you a better sense of proportion.
5. Talk over major grievances calmly. Try writing down the items that irritate you on slips of paper as they occur. Say nothing at the time. Later, when you and your husband are both calm and serene, take out the slips and look them over. You will be ashamed to mention the trivial and unimportant grievances and will throw them away; but discuss the major causes for irritation reasonably and unemotionally.

Recently, at a banquet, I was seated next to the manager of industrial relations of one of the oldest companies in the U.S. I asked him for his ideas on how wives can help their husbands get ahead.

"I believe," said this executive, "that the two biggest things a wife can do to help advance her husband's career are (1) love him and (2) let him alone. A loving wife will see that her husband has a comfortable, happy home life. And if she is smart enough to let him attend to his business without interference, there is no reason why he can't advance as far as his ability and training will take him.

"A wife can literally meddle her husband right off the payroll," he told me, "by advising, by interfering, by influencing him against people he works with, by complaining about his pay, his hours and his duties."

Many brides have rosy dreams of subtly maneuvering their dream-boys right up into Executive Row. In case you are one of the girls who believes in wielding behind-the-scenes power, I'll make it easier for you: Here is a list of ten techniques by which you can hamstring your husband and drag him down the ladder, instead of helping him up.

1. Be nasty to his secretary, especially if she is young and pretty. Never pass up a chance to put her in her place. Losing a good secretary can be a major disaster to an ambitious man.
2. Phone your husband several times a day. Tell him your domestic troubles, ask him whom he's lunching with, and don't forget to give him a list of groceries to pick up on his way home. Never fail to meet him at the office on payday. His co-workers will soon find out who is boss at your house. And his powers of concentration on his work will vanish.
3. Start a feud with the wife of one of his associates. Soon the whole office will be divided into factions—and it won't be long now!
4. Tell him how overworked and underpaid he is and how nobody appreciates him in that office. Sooner or later, he may begin to believe you and his work will show it.
5. Make a habit of telling him how he could do his job better and curry favor with his superiors. After all, he only works at the office—you're the master-planner.
6. Give him an air of success by throwing expensive parties and living beyond his income. You'll fool nobody, but you'll have lots of fun—while it lasts.
7. Organize your own home spy service by cross-examining him constantly about his relations with female clients, office help and wives of associates. The fact that women are in business to stay and a man can only escape dealing with them by setting up shop in the Men's Room means nothing to you. You know they're all scheming hussies.
8. Use your sex appeal every time you get a chance to make eyes at his boss. If the boss doesn't give him the axe after this, the boss's wife will see that he gets a new boss.
9. Drink too much at office parties and conventions. You'll be the life of the party—and provide, at his expense, endless gag material for the folks he works with.
10. Cry, complain and nag every time he has to work overtime or go on a business trip. Make him realize that you come first.

Follow these ten rules, if you want to do a first-class job of fouling up your husband's opportunities for promotion. Chances are, he'll lose his job and you'll lose your husband.

Is your husband prepared for promotion? If not, what is he doing about it? Very few men have the knowledge, at the start of their careers, that they will need for the jobs they hope to get 5, 10 or 15 years later. They must learn as they go along, both by experience and by special training.

Sociologist W. Lloyd Warner says that the American dream is based on the belief that a man can "get ahead" — and one of the main ways by which a man moves upward is education. Many firms provide special training programs, at company expense, for employees. Others award promotion to men who have the initiative to take special training on their own time at their own expense.

What part does a wife play in a man's efforts to educate himself for promotion? Mainly this: her attitude will affect his efforts to improve himself.

Take the matter of night-school training, for instance. The man who devotes two to five nights a week to night-school classes is undoubtedly eager to forge forward, either in his present work or some other field for which he is preparing himself. His wife must learn to get along without him during this period. She must adjust herself to hours of loneliness and fill up the gap with activities of her own.

If she fails to make this adjustment, part of the man's necessary concentration on his studies will be clouded by uneasiness over his wife's unhappiness. Sometimes he gives up his educational endeavors because of her complaints about being left alone.

The wisest course for the wife is to line up a study program of her own. Perhaps, if finances permit, she can take the same training as her husband to give her a more intelligent grasp of his work; or she can study some allied field and supplement his knowledge.

The wife who wants her husband to succeed must be willing to let him work at whatever he loves best, even if it means taking risks. She must have the courage of his convictions and not be afraid to back him up in his ambitions, regardless of where the chips fall. Those who sacrifice initiative and enterprise to security are likely to wind up with nothing else.

I know a man who is serving a life sentence at uncongenial employment. He started as a bookkeeper to earn money enough to open his own auto repair shop. Then he got married; and his wife thought he should keep his job until they had saved enough for a down-payment on a home.

When that goal was reached, a baby was on the way. This man's wife made him see how foolhardy it would be to risk their meager savings to start his own business — and the years went by.

There were payments to meet on their home — insurance to keep up — their son's education. Start out on his own? Ridiculous! What if he didn't succeed? He would have lost his seniority in the company, the firm's old-age pension, sick-benefits and a steady, if modest, salary. So this man lost his chance because his wife wasn't willing to take a chance on him.

Today, he is a tired, bored, middle-aged duffer who spends his spare-time tinkering with his car. He has a beaten look and nothing much to remember. Life has somehow passed him by.

What if he had given up his job, tried his hand at his chosen work and failed? At least he would have had the satisfaction of having reached for the thing he desired. And if he had tried and failed often enough, he might eventually have succeeded.

If we want our men to succeed in the work that offers them the greatest fulfillment, let's encourage them to take a chance—and be brave enough to share the risks.

Somewhere in a New Zealand cemetery, writes author E. J. Hardy, is an old gravestone bearing a woman's name and the words: "She was so pleasant."

I don't know how that affects you, but, personally, I can't think of an epitaph I would rather deserve. The grieving husband who put that on his wife's tombstone was blessed with a thousand memories: a face lit up with smiles at his return—hot meals on the table—someone laughing at his worn little jokes—a home that closed him in with love and comfort.

Notably successful marriages are built on a wife's thoughtfulness in learning and doing what will please. Mrs. Dwight D. Eisenhower says that she considers it a woman's primary job to remember the little things which contribute to the happiness of others.

Maybe these little things aren't so little after all. Wasn't it Lord Chesterfield who said that "good manners are made up of petty sacrifices"? That is also the secret of good marriages. Wives who are willing to give up some of their own preferences are usually rewarded out of all proportion to the petty sacrifice.

Sharing anything—be it a crust of bread or an idea—brings people closer together. Sharing the special interests and recreations of those we love is one of the main highways to happiness in human relations.

What are the basic elements of companionship? Common friends, common interests and common ideals—these are the things that bind people together.

Arthur Murray and his wife Kathryn have probably taught more people to dance than have any other two instructors since the beginning of time. The Murrays have been married for 28 years and have worked together as partners all that time.

I asked Kathryn Murray: "Working as closely with your husband as you do, how do you keep from getting in a rut? Don't you find it hard to separate your business life from your life as husband and wife?" "Not at all," said Mrs. Murray. "It's merely a matter of my making a little extra effort. I try to dress attractively at home for one thing—and I would rather have ten men see me without powder on my nose than for my husband to.

"But, more important, we share similar interests. We both like swimming and tennis. Whenever we can, we take vacations together and enjoy these sports. Last week we had a quick trip to Bermuda. Sharing our fun brings us together on a different basis and adds variety and zest to our life."

How many of us will put on hipboots and dungarees, get wet, dirty and cold, and bait our own hooks just to be companionable with our husbands?

A highly eligible bachelor confessed to me that he would marry like a shot if he could find a woman who would give him companionship and, at the same time, have respect for the fundamental male urge to be left alone when he feels like it.

Housewives spend so much time alone they often fail to understand that a man's being "left alone" does not imply real loneliness—it just means being set free from all female demands and constraint. Some husbands achieve this illusion by taking a night off to bowl or play pinochle with the boys. Others shut themselves up in the garage and overhaul the car—or read a detective story. Whatever specific use a man makes of these happy moments of aloneness, it's smart for a wife to see that he gets them.

No doubt about it, husbands need to slip the leash occasionally. If we can aid them to follow up some absorbing, sparetime hobby—and also give them a reasonable measure of utter freedom—then we are doing a lot to make them happy.

Another way to be a good companion is for a wife to have some separate, outside interests of her own. Just as a man goes back to his job strengthened and renewed by a few minutes or hours spent at an interesting hobby, so does a wife approach her duties in a better frame of mind when she has some outside activity. It's the change of activity that refreshes.

Sparetime activities which bring wives into contact with others are most beneficial. A course in consumer education or millinery, a music-appreciation class, a few hours a week working with some charitable or civic organization—projects like these give a woman a fresh viewpoint and make her more of an individual.

Look inside yourself—think of what you have always enjoyed or wanted to do. It needn't cost money. Look over your community—you will be amazed to discover how much worthwhile (and inexpensive) activity is offered by even the smallest towns. If you can't find what you want, get busy and organize a group of other people who want the same thing.

What kind of an atmosphere does your husband come home to after a busy day? And what kind of a home springboards him to work and renewed effort every morning? The answers to these questions may have more to do with his success—or lack of it—than you think.

To enable a man to work at top efficiency, his home must provide him with certain basic elements:

1. *Relaxation.* No matter how much a man likes his job, a certain amount of tension is built up in his working hours. If this tension is broken when he goes home, a man can re-charge his mental, physical and emotional batteries.

Every woman wants to be a good housekeeper, but sometimes a man finds no relaxation at home because his wife is too good a housekeeper. All of us wives have an occasional impulse to use a blunt instrument on our mates when they strew Sunday papers, cigar stubs, empty glasses and assorted items over the carefully arranged, inviting house we have worked so hard to get that way. But before sounding off about what an inconsiderate bum he is, let's remember that home is the only place where he can relax and be his sloppy, lovable self.

2. *Comfort.* Since decorating and furnishing the home is largely done by the wife, she must remember that comfort is a man's major requirement. Spindly tables and chairs and clutters of knickknacks may charm the feminine eye, but they spell nuisance to a tired male.

3. *Order and cleanliness.* Meals that are rarely on time—litter in the bathroom—unmade beds; these and other signs of unfinished business in the housekeeping department can drive a man to poolrooms, saloons and blondes. For men, funny critters, can't seem to endure anybody's messiness but their own.

The impression other people have of your husband is quite frequently a reflection of your own attitude towards him.

Not long ago, I called up a local appliance dealer to inquire about an electric cooling system. The dealer's wife took my call and gave me the information I wanted. Then she said: "Of course, Mrs. Carnegie, my husband is the real expert on cooling systems and if you will let me make an appointment for him to look at your house, he can then recommend exactly the type of fan you need. I can only guess, but he knows."

When the man came to check over my house, I was already prejudiced in his favor by his wife's confidence — all he had to do was follow through and make the sale.

People tend to live up to the character we give them. Tell a child he is awkward and he will be clumsier than ever. Praise him for politeness and his manners will improve. Treat a man as if he were successful and, unconsciously, he will begin to display the qualities that make for success.

Wives of professional men seem especially adept at creating favorable impressions of their husbands' ability. "I wish we could go to the party," they tell you sadly. "But Bill is snowed under right now preparing his brief for that big Jones Company lawsuit."

In a few offhand words, these girls create a mental image of their boys as up-and-coming lads who have to fight off clients (or patients) with a bat in order to find time to breathe.

No modest man likes to blow his own horn — but it does no harm for his wife to give it a few toots, provided she keeps within the bounds of good taste.

On the other hand, a man can be too modest for his own good. If your husband is one of those who habitually make light of their own accomplishments, there is danger that others may eventually take him seriously and decide that he really isn't such a ball of fire after all.

Most men eat more and need less food as they grow older, because they are less physically active. It's our business to establish good eating habits early in the game, if we want to keep our husbands' weight down and their spirits up.

See that the meals your husband eats at home are free of haste and tension. The morning breakfast dash is sadly familiar in too many homes. Get up earlier, if necessary, to see that your husband gets, at least, a leisurely, nourishing breakfast.

Here are some rules to follow if you want your husband to live longer and feel better:

1. Watch his weight as carefully as you do your own. Write any insurance company and ask for a weight-longevity chart. Check your husband's weight against this chart and see if he is as much as ten per cent overweight. If he is, ask your family doctor to prescribe a diet.

2. Insist on annual medical, dental and optical checkups. Many deaths from heart disease, cancer, tuberculosis and diabetes could be prevented if they were discovered in the early stages.

Overambition may make him successful, but he isn't apt to live long enough to enjoy it. Develop courage to influence him to turn down promotion when it means too much added strain and overwork.

3. The secret of resisting fatigue is to rest before you get tired. Short periods of relaxation work wonders. If your husband comes home for lunch, get him to lie down for 10 or 15 minutes before he returns to work. Encourage him to take short naps before dinner. It may add years to his life.

4. Keep his home life happy. An unhappy, worried or angry man is "accident prone"—so keyed up inside that his reflexes don't work properly. He is likely to wreck his car on the highway, or get fouled up in the machinery if he does mechanical work.

He is also more likely to eat or drink too much. Dr. Harry Gold, of Cornell University, says that "people often take to eating when they are unhappy, or to gain release from depression or tension."

A big part of everybody's success in life is being healthy enough to enjoy it. And whether we wives like it or not, we must accept responsibility for our husbands' health too. "My Life Is in Your Hands" could very well be any married man's theme song.

24

PAUL H. LANDIS[1]

"What Is 'Normal' Married Love?"

Coronet, October 1957

By trying to adhere to almost impossible standards, many couples cause themselves great unhappiness.

The neat and attractive young woman was ill at ease as she sat down before the marriage counselor's desk. "I don't know what's wrong," she said desperately. "I should have a perfect marriage. I have everything I

---

[1]Paul H. Landis was a professor of rural sociology at the State College of Washington (now Washington State University). He also wrote books about marriage and family life, including *Your Marriage and Family Living* (New York: McGraw Hill, 1946), *Understanding Teenagers* (New York: Appleton-Century-Crofts, 1955), *Making the Most of Marriage* (New York: Appleton-Century-Crofts, 1955), and *For Husbands and Wives: A Plan for Happy Marriage and Family Living* (New York: Appleton-Century-Crofts, 1956).

Paul H. Landis, "What Is 'Normal' Married Love?" *Coronet*, October 1957, 126.

want. My husband and I love each other and we have two wonderful children. Yet I feel I am missing out on the most important thing of all."

As the discussion continued, it became apparent to the counselor that Janet really did love her husband deeply, was content in her day-to-day life and could truthfully say that in general she had a happy marriage. Yet it was just as obvious that there was an undercurrent of anxiety and discontent within her that was threatening to destroy what she had built and wipe out the compatibility and love she shared with her husband.

Finally Janet admitted the reason for her unrest and concern. She had read much of the flood of material now available on sex problems; and had continually encountered the so-called "normal" standard, otherwise described as "sexual perfection," "sexual harmony," or the "perfect sex life." Comparing her own experiences to those described, she could only conclude that she and her husband were not achieving the modern sexual goal. As a result she felt inadequate when it came to pleasing her husband and at the same time felt cheated and irritated with what she suspected was his ineptness.

Janet's problem is not an unusual one. In fact, her worries are typical of thousands of today's married couples. Failing to live up to the "norm" they hear so much about, they conclude there is something wrong. The result is agonizing conflict and, often, a broken marriage. The great tragedy is that many of the marriages that fail as the result of sexual difficulties would have a good chance of being saved if the couples understood more about the so-called sexual "norm" and "standard."

To begin with it is necessary to know what is meant by the term "norm." A norm is based on the average of what people in general are and how they act. It is not, contrary to popular opinion, "a standard to be used to judge the nature and actions of individuals," or, in this case, of individual married couples.

Much of the confusion about a standard of sex performance has grown out of the great change of attitude toward the wife's role. In the course of two or three generations we have shifted to a position directly opposite that held by earlier generations. Sex pleasure was a sin then, even in marriage. The catering of the wife to the "animal" nature of her husband was a necessity, a duty performed in exchange for the security of her home and children.

Previously, too, the main purpose of intercourse from the woman's point of view was to have children. Now sexual experience is also regarded as an expression of mutual love and unity. Women are expected to participate in the pleasure and are allowed to show desire and active response without a feeling of guilt and without exposing themselves to censure.

Yet, this new attitude, which has brought so much joy, has also brought with it a great deal of unhappiness. The place of woman in sexual activity has been "discovered," but not accurately defined. They know they are supposed to take a greater part, but they don't know how great a part. They know that they are supposed to get more satisfaction out of sex, but they don't know how much.

The crux of the problem involves the sexual climax. Wives read, or they are told, that the "complete" sexual performance includes orgasm. If they do not experience this, as in Janet's case, they begin to feel that they are a failure as a wife, that they are frigid or inadequate. Often, too, they begin to feel cheated and resentful. They begin to doubt their husband's ability as a lover. They feel wronged because they are not gaining as much pleasure from the experience as he apparently does.

This unhappy situation is the result of a myriad of misunderstandings. Take the idea that husbands and wives are supposed to react with equal intensity at any given time. On the basis of scientific fact, this is not only improbable but, in most instances, impossible.

In the first place there is a vast difference in the sexual development of men and women. Since the male has his most vigorous period in his late teens and early twenties, and the female's desire does not begin to reach its peak until the thirties, the husband is likely to exceed his wife's desire in the early years of marriage. This variance in response leads to trouble when the wife begins to wonder why she does not react as vigorously as her husband. And the husband begins to wonder whether his wife really loves him.

Too, from a physical point of view a woman's response is not as localized and as immediate as a man's. Her responses are spread out over her body and are, therefore, more diffuse, and appear slower and less intense. There are other facts that add to the imbalance of the sexual relationship in early years of marriage. During the twenties the wife is likely to be bearing children. With the burden of pregnancy and caring for small children, the wife's sexual interest is further subdued and interrupted. The gap between her reactions and those of her husband is made even greater. Her failure to live up to the so-called "norm" is even more obvious.

For a woman, sex is just one incident in a whole sequence of events — pregnancy, childbirth, lactation, and child-care. She, therefore, cannot be expected, despite the modern emphasis on her participation, to have a psychological attitude toward sex exactly similar to that of her husband.

Nor can a woman expect to respond to sexual experience in the same manner as some other woman might. Some reach a climax regularly. Some less often. Some not at all. For different women have different physical and psychological make-ups. Some are more restrained or more inhibited than others as the result of training and experience. Some are more physically able to respond because they are naturally more sensitive to stimulation.

Research in this country and my inquiries into clinical findings in European and Scandinavian countries, indicate that ten per cent to a third of women rarely or never arrive at a climax in sexual relations. Kinsey's research shows that only three-quarters of the sexual experience in American marriages produces a climax for the wife. This ratio is much less during the early years of marriage and higher during the later years. In the first year of marriage, only sixty-three per cent of the Kinsey women reached this goal, whereas in the twentieth year of marriage, eighty-five per cent did.

Anthropologist Margaret Mead found in her study of various societies in the Pacific Islands that some whole cultures discourage women from being assertive in sexual relations. All that is expected is that they be receptive. There is nothing in the study of comparative cultures, Dr. Mead concludes, to lead to the assumption that orgasm is an integral and unlearned part of woman's sexual response as it is of a man's sexual response. We do violence to woman's nature to assume that, for her, regular orgasm is natural.

The exaggeration of the physical rewards of sex has had serious side effects beyond causing personal doubts and unrest. It has led some women to try futile experiments outside of marriage in the search for a thrill they never get. The late Dr. Lewis M. Terman found that women who do not respond in marriage seek outside liaisons more often than women who do respond. But he also found that these stories usually fail to produce the kind of results the wandering wives seek.

The male, too, is paying a high price for the spurious emphasis on physical aspects of relations between the sexes. The psychological demands of his sexual powers are greater than they have ever been before. Under the old scheme of values, he initiated the sex act and performed it to his satisfaction. There was no occasion for him to question his adequacy. Now, aware that he is supposed to satisfy his wife, and bring her to a climax, he worries if he fails to achieve this. As a result the weaker-sexed male loses the confidence and readiness he needs for an effective and satisfactory relationship.

As Margaret Mead also points out in her book *Male and Female,* the current exaggerated concern over sexual response of the mate dampens the spontaniety of both partners. Women, now that they know what sex experience means to men, often worry for fear that they are not giving their husbands adequate gratification. Men, now that they understand the way a woman can respond, "worry as to whether or not their wives are unsatisfied." These twin worries could well make couples unable to "respond simply and immediately to each other," Dr. Mead warns.

In my own research, comparing the viewpoints of married girls of college age with the attitudes of their mothers, I have found that the current generation is more concerned with sex problems than their mothers were. Certainly they know much more about the subject. They often discuss plans for having children with their prospective husbands before marriage, and are given far better factual preparation for marriage through better sex education.

Yet, even though today's wives know more about sex than their mothers did and are likely gaining more sexual satisfaction than their mothers did, they are not as happy as they should be because they are led to expect too much and are urged to strive for goals they cannot reach.

It must be realized that sex cannot be appraised solely as a sensuous experience for either husband or wife. The degree of contentment and relaxation it brings, the relief from tension, the feeling of unity, the comfort and sense of belonging, count more in the long run. If supreme peaks of sexual excitement are followed by release for the wife, so much the better. But if she is happy in the relationship, relieved and content, neither she nor her husband need worry about her failure to experience the kind of reaction others may experience.

Marriage is a complicated interrelationship involving many things. Sex is just one of these. If a marriage fails on other points, a strong sexual attraction or physical compatibility will not save it. A couple who quarrel bitterly much of the time, yet are drawn to each other by sexual need do not have a successful marriage. Their failure will take its toll on each partner in the marriage, and on their children. But a couple who love each other, whose day-to-day life is satisfying and tranquil, and who are building a happy future together can be counted successful. The wife's failing to achieve a climax during the sex act may well be disregarded.

After all, working out a satisfactory sexual adjustment in marriage takes time. Dr. Judson T. Landis, Professor of Family Sociology at the University of California, studied the length of time required by over 400 married couples to achieve adjustment in their sexual relationships. Only half were mutually adjusted from the beginning, and one in eight had not adjusted after more than 20 years. Yet these couples found other areas of their marriages sufficiently rewarding to offset this one disadvantage, since they had been married an average of 20 years or more and had reared children.

To get back to Janet in the counselor's office. She is not, as she mistakenly believes, "missing the most important thing." She is young enough so that perhaps her sex response will in time improve, and is straightforward enough to be able to suggest to her husband that he could improve his approach. Perhaps a little later in her married life she will reach the modern sexual goal. However, if she does not, indeed if she never does, her marriage can still be a happy one. The only thing marring her content is her mistaken feeling that she is not measuring up to today's standard for a wife. The sooner she and all others who share this misconception realize that it is a *false* standard, the better it will be for their marriages.

25

BERYL PFIZER

"Six Rude Answers to One Rude Question"

McCall's, July 1960

There is no place on earth where a single woman is safe from the rudest, least answerable question ever asked: "Why aren't you married?" At cocktail parties, art exhibits, class reunions, business lunches, political rallies, and clambakes, the unmarried girl can always count on some well-meaning busybody's coming up with the question. Ten seconds after you've met them, normally polite people, who wouldn't dream of asking you your age or how much salary you make or if you had an unhappy childhood, ask you to explain your single status. It is as if, ever since the animals clambered onto the ark two by two, there's been something subversive about a loner.

Oh, sometimes the question is phrased differently. Frequent variations are: "How come you're not married?" "When are you going to break down and get married?" And even the fatuous "Why isn't an attractive girl like you married yet?" However it's asked, it's rude. And it's time the Single Girls of America united to put a stop to it.

Before I set down a suggested list of question-stoppers, however, it's

Beryl Pfizer, "Six Rude Answers to One Rude Question," *McCall's*, July 1960, 44.

important to note a few exceptions. There are some people who, for various reasons, cannot or *should* not be stopped. Mothers and doting great-aunts, for instance, are expected to ask the question. It's part of their nature. These, however, can be put off gently but firmly by clever use of some inside family information. Just ask casually, "How old was Aunt Blanche when she married Uncle Charley?" (You know she was thirty-nine, and it was a *very good* match!)

Another group to be excepted are comparative strangers—dentists, salesmen, and elevator operators. Even if you had a perfect answer, you couldn't deliver it to a dentist with his drill in your mouth. As for the salesman, he just doesn't get an order. The elevator operator finds a little something less in his Christmas envelope, and after a while, you get used to that long wait for the elevator.

Butchers are in a class by themselves. They must be jollied along with a smile and a gay little laugh—unless you happen to *like* tough meat.

All right. So much for the exceptions. On to the real offenders and how to handle them. Emily Post[1] offers no help. But here are the results of years of experimentation and research: six tried-and-true answers to the question "Why aren't you married?"

1. When the questioner is a married woman you've known for some time and you detect a slightly catty tone in her voice, pull off your kid gloves. Look her straight in the eye and say, "I'm not ready to lose my looks *quite* yet."

2. When the questioner is a married woman you don't know well, and don't care to, try this: "Confidentially, I prefer having affairs." Then let your gaze wander to the man you know is her husband and ask innocently, "By the way, who's that attractive man in the gray suit?"

3. If you have a flair for the dramatic, you might give the girls something to whisper about in the powder room and add a little glamour to your reputation this way: Lower your eyes, sigh, and say, "I just can't bear to talk about it." Squeeze out a couple of tears if you can, but not enough to ruin your eye make-up.

4. An unmarried friend of mine has great success with: "Oh, it's a long story. It takes about three hours, but I just love to tell it. Let's go sit down somewhere comfortable." The questioner invariably remembers she was just leaving.

[1]Emily Post (1873–1960) was the national authority on etiquette and good manners. Her book *Etiquette in Society, in Business, in Politics, and at Home* (New York: Funk and Wagnalls, 1922) became the bible of etiquette for generations of Americans. She also wrote a syndicated daily newspaper column and hosted a nationwide radio program.

5. If you're the type who likes to get a laugh, or if there's a man listening and you want him to think you're a great little sport, keep this old joke handy: "Oh, I could have married anyone I pleased. I just haven't found anyone I pleased."

6. For the really alert single girl, here's the best answer. It depends entirely on timing. Let the questioner get as far as "Why aren't you marr—" Then spill your coffee on her. She'll lose interest in your answer right away.

Those are six sure-fire methods of handling the problem. At least, they've worked for me. As a matter of fact, after practicing them for some time, I find the problem has been eliminated. No one has asked me any*thing* or any*where* in months.

One present you can use right away...

new SINGER "Roll-a-Magic" vacuum cleaner

The SINGER "Roll-A-Magic's" *whirlwind* suction will whisk up Christmas clutter instantly. This new roller cleaner has *double* the capacity of most cleaners—cuts bag replacements in half.

And it swivels both top and bottom to reach every part of room without effort. See it today at your nearby SINGER SEWING CENTER . . . or call for free home demonstration.

$79⁹⁵
complete with attachments
Only $1.25 weekly after minimum down payment

Now, SINGER offers all 3 types of cleaners

SINGER "Magic Carpet" Cleaner
Only cleaner made with *double fan* suction . . . the *finest* ever made for rug and carpet cleaning. Hangs up flat against closet wall . . . automatic cord rewind. $89.95. As little as $1.25 weekly after minimum down payment.

SINGER "Magic Mite" Hand Cleaner
Largest-selling hand cleaner. Perfect for stairs, autos and furniture. $25.95. Only $5.00 down, balance on easy terms.

Here's why a SINGER® is your best vacuum buy:

- Each cleaner is the finest, most efficient of its kind.
- A SINGER Cleaning Consultant is as near as your phone.
- Immediate service from your nearby SINGER SEWING CENTER. (Phone number listed under SINGER SEWING MACHINE CO.)
- Low weekly terms without embarrassing red tape.

 SINGER SEWING CENTERS
(A TRADEMARK OF THE SINGER MANUFACTURING COMPANY)

Figure 4. Singer advertisement from *Woman's Home Companion,* December 1956, 70.

4

Homemaking

In April of 1940, the *Ladies' Home Journal* reported that 94 percent of American homes had electricity and that 95 percent of American women did their own housework—that is, without the help of servants. In a way, these statistics seem to bear out the prophecy of Thomas A. Edison, who, in "The Woman of the Future," published in *Good Housekeeping* in 1912, foresaw the day when the housewife would be "neither a slave to servants nor herself a drudge," but instead a "domestic engineer" assisted by "the greatest of all handmaidens, electricity." In Edison's rosy prediction, women would use the time that electrical appliances saved them to pursue intellectual activities that would allow their brains to evolve, finally, to equal those of men. In reality, the numerous household appliances available by midcentury merely raised the standards by which housekeeping was to be judged. Margaret Halsey, in *This Demi-Paradise,* her 1960 account of life in the suburbs, recalls the "Invisible Critic" that she imagined hovering near the ceiling of her home, who never delivered the praise for which she continually hoped.

In many ways the women's magazines reinforced the search for housekeeping perfection. Typical is an article in the April 1950 issue of *Good Housekeeping* that equates a pretty, efficiently organized kitchen with the skills necessary to make "a perfect lemon pie." Yet it is also clear from the magazines that women needed a great deal of help to become "domestic engineers," and they set out to provide it: lists of utensils that a bride needed to set up her first kitchen, menus and schedules for weekend entertaining, and articles on how to dust, shop, and budget time and money. The Good Housekeeping Institute advised homemakers most scientifically, testing products ranging from lipsticks to washing machines; most other magazines were somewhat more casual but no less earnest, promising "Better Meals with Less Work" and freedom from "wash day drudgery" with the proper appliances. Some women, such as the mother of four featured in "The Scrambled Housewife" in the August 1954 *Journal,* needed more help than others. However, housekeeping was to be, ideally, joyful and rewarding. Such is consistently the tone of Gladys Taber's "Diary of Domesticity" feature in the *Journal,* with its cozy chatter about cooking and puppies. Further, homemaking was not merely a set of responsibilities but a "profession" or an "occupation," as Dorothy Thompson (herself a professional journalist) in 1949 reminded an acquaintance who felt she had "wasted" a college education by remaining a homemaker. A decade later, the voices of such women were finding their way into Betty Friedan's *The Feminine Mystique.*

Two types of articles on the homemaker's role appeared in the magazines. The most common sort of article was practical and to the point: Using a "how-to" format, usually with both text and illustrations, these articles provided instructions to the reader—recipes not merely for food but for a host of household tasks. The second kind of article debated the role of homemaker itself: Was she truly a "professional," as had been argued since the previous century, deserving status and respect, or was she simply isolated and unfulfilled, a slave to unrealistic standards in the performance of her job? Friedan took the latter position in her 1963 book, holding the women's magazines partially responsible, but a careful reading of the debate as it was carried on in the magazines suggests that the reality was far more complex.

26

GRACE L. PENNOCK

"Starting from Scratch"

Ladies' Home Journal, April 1940

American women have the most modern kitchens in the world, more laborsaving housekeeping aids than those of any other nation. To do their ironing, 94 per cent of U.S. housewives in homes with electricity (23,420,000 out of 25,204,976) use electric irons; more than half—57 per cent—own electric washers; and 51 per cent have electric refrigerators. As a result, 95 per cent of all America's housewives do [their] own housework, find the job smooth-running, zestful, exactly to their taste.

Grace L. Pennock, "Starting from Scratch," *Ladies' Home Journal,* April 1940, 74.

Add to one small furnished apartment a bride with no experience in cooking and housekeeping, and you get either a great adventure or a very bad situation. It's exactly the setup Ginger faced when she started housekeeping. Thanks to her good sense and ingenuity, it has turned out to be a great adventure.

Her mother headed her right by telling her to use "common sense, good judgment and patience" and she could learn to cook, sew and keep house as well as anyone. By using her head, Ginger finds housekeeping both fun and a very serious matter.

"Yes, I have a schedule." She plans her work so she will have the house in order, some cooking done ahead, and free time herself over the week end when Ted is home. She does a "good going over" of the house on Thursday. The linoleum floor in the kitchen she leaves until Friday because "I seem to be pretty sloppy." Anyone would, in such a small kitchen.

"Ted tells me to go by the cookbook." Ted coaches her in using the cookbook. When she asks him what to cook, Ted usually says, "Try something different." They shop Saturday afternoon. "We want to get our money's worth out of everything," she tells us. They believe in enjoying their best things by using them. If they're good, they will last a lifetime anyway, is their way of looking at it.

"I had such fun." Like any woman, new or old at this job of homemaking, she gets lots of satisfaction in fixing things up. "It has really been so much fun trying out different ideas." As to arranging dishes on her shelves, she said, "The shelves are so attractive now. The space was small and I played checkers with dishes for three months before I got them to fit."

"When I get in a jam." When she doesn't know what to do about some puzzling matter, she just sits and thinks. If it all gets in too big a jam, she leaves it and takes a walk. She usually thinks of an answer. Not such a bad plan for any housekeeper, new or old.

Plan Ahead

We asked Ginger if she ever kept Ted waiting because she forgot to put in the roast on time. Did the peas ever cook to mush while she made gravy?

If you are a young bride, such catastrophes often happen unless you plan carefully.

Plan *simple* menus. Never try more than one unfamiliar dish per meal. Don't plan more top-stove dishes than you have burners. See that oven dishes bake at same temperature. Check refrigerator and supply cupboard before you shop. The Chases do most of their buying on Saturdays and keep makings of a meal or two on hand at all times.

Prepare Ahead

With the menus planned and the marketing done, we told Ginger it would help her considerably to jot down on a pad just about how long it will take to prepare and cook each dish—as well as which foods can be partly or completely prepared ahead of time. These beforehand chores may then be dovetailed with other duties.

For instance, in Ginger's dinner plans the custards can be baked, sweet potatoes scrubbed, broccoli washed and pared and the fruit juice chilled—all before lunch.

While you're in the kitchen, some preparation may even be done toward next day's meals.

Mealtime Roundup

Ted comes home about 5:30 P.M., so 4:30 is Ginger's deadline for beginning the evening meal. She sets the table first so she's sure not to forget things like filling the salt and pepper shakers. From then on it's simply a matter of minding her peas and cues.

For instance, when the sweet potatoes have been baking for about thirty minutes it's Ginger's cue to put on the broccoli—and so on till dishing up when Ted comes home.

As she grows in experience she won't need to make a schedule. Tasks will follow one another naturally. Ted and Ginger agree it's a good idea when first beginning—if you don't make too hard work of it.

27

DOROTHY CANFIELD FISHER[1]

"Housekeeping Need Not Be Dull"

Ladies' Home Journal, October 1941

"How does a woman in a home find the time to keep on growing mentally?" was the question Jean Bevan Abernethy asked me. She wrote, "We are amazed and appalled at the way in which our day seems to dissolve before our very eyes. It is frittered away, and the worth-while things we thought we were going to accomplish are not done. We are no longer protected, as we were in college, from the talkative delivery boys, the telephone calls, the planning of menus, the washing of dishes, shopping, bridge, committees, dining out and giving dinners. Trifles and routine engulf us. Then we say 'Tomorrow'; and tomorrow is like today.

"There must be a way to keep us from becoming dull and insensible—from 'getting into a rut' which we foresaw in college—a way that is concrete and practical and comprehensible.

"We do not want a neat set of ten commandments in which we can bury our conscience and forget about it. We do want and need desperately the methods which honest women have found to work—we want practical suggestions which have helped at least one person, and which may help us. You who are older, you who were engulfed in routine as we are, and managed to escape—how did you do it?"

Well, here is my answer:

First of all, well-educated and intellectually alive young housemother, remember that you can't have your cake and eat it too, in homemaking any more than anywhere else. Living is an art. Living as a wife and mother is a great art. One of the first principles of art is the elimination of the unnecessary, leaving things out, even when they are pretty, or fine, or

---

[1]Dorothy Canfield Fisher (1879–1958) had a varied career as a translator (she received her Ph.D. in French from Columbia University in 1905), novelist, advocate of the Montessori method of education, and member of the editorial board of the Book-of-the-Month Club (1926–1951). Her books include *The Squirrel Cage* (New York: Henry Holt, 1912), *The Bent Twig* (New York: Henry Holt, 1915), *Fables for Parents* (New York: Harcourt, Brace, 1937), and *The Montessori Manual* (Chicago: The Richardson Company, 1913).

Dorothy Canfield Fisher, "Housekeeping Need Not Be Dull," *Ladies' Home Journal*, October 1941, 156.

attractive things in themselves. Which for you is not necessary—bridge or Beethoven?

Another basic principle of art, and hence of living, is constant prayerful attention to the general proportions of what is being created by the artist. You'll often see a beginner in a drawing class carefully shading the convolutions of an ear in a head which is grotesquely out of proportion. But art students have professors who tell them, "For heaven's sake, look at the size of that head compared with the ear. Scrunch that sheet of paper up, take a fresh one, and this time concentrate on getting the whole thing in true proportion."

These are the great basic principles which you yourself must learn to apply to your own life. Here now are some rules of thumb which I have found useful.

Buying things is a considerable part of your work as housekeeper. Arrange your buying so that you are not always at it, above all so that you do not think about it a single instant longer than necessary. Take an evening off for planning, consider with your husband seriously what you must buy, for as long ahead as you can conveniently plan—decide what you have to spend, go ahead and do the buying—and then dismiss the whole matter of purchases from your mind for days, for weeks, for months. Do not ever step into a shop "just to look around."

In the matter of food, buy the staples in as large quantities as practical, so that you don't have to think about them daily; have simple meals; give half an hour in the morning—or better, in the evening—of intensive thought to planning food for the next day. Then don't give any more real thought to the matter.

Here is a good place to put in a warning about a well-hidden ambush from which an unsuspected attack is launched on everybody who tries to make his life something more than a trivial round of picayune material details. We say, and sincerely believe, many of us, that we waste time in fussing over unimportant details because we will be criticized if we don't, because our husband's clients or patients or students or employer will think less of him, or our children's friends look down on them, if we don't come up to a certain standard. Sometimes this is partly true. But vastly more often we give our time to fussing over those things because it is easier and (momentarily) more pleasant than to stick patiently at the long, slow struggle to attain true excellence in something that is worth while.

Do not think that homemaking women have any monopoly of this flinching from real effort. It is a universal human failing. All people who—

like women in their homes—work without foreman, supervisor or time clock, must perpetually fight this escape instinct to waste their time with easy little processes which can be counted on to give one quick little satisfaction after another.

Lastly, I think that American young women of the kind represented by Mrs. Abernethy would do well to take a leaf out of European practice and do less of the housework themselves. Part-time help in housework costs less than many possessions which many people have, and which a European family of cultivation would get along without, in order to save its homemaker some time for music, reading, study or professional work. Most of our young American couples of education and cultivation—except in times of great commercial depressions—are twice as well dressed, ten times as well shod, forty times as well provided with plumbing as Europeans of their kind—and hire a tenth as much help in the house and have vastly less time for music and good books.

American women, with their over-emphasis on material details, are considerably to blame for this practice. You hear them say by the thousands, "Oh, but the help you hire nowadays never do it right." What of it? Cooking is the only part of housework which it is really important to do just right, and you can keep the cooking in your own hands. As to other things—which is more important, having the front porch and the kitchen floor as clean as you could keep them yourself, or having time—not for playing cards or more fussing over table linen—for Plato, or international affairs, or economics, or reading aloud to the children?

But even if this simple device is not possible, let the intelligent young homemaker remember that not all her life is going to be like these beginnings. Let her remember that the profession she is now learning will be easier to practice when she has learned it, than she now thinks possible. If things go normally for her and her husband, in no time she will see younger women looking enviously at her and asking, "How do you do it?"

28
DR. CARL P. SHERWIN[1]
"The Question-Box"
Good Housekeeping, January 1944

Is arsenic used in the manufacture of commercial vegetable shortenings?

Mrs. F. D. P.

There is no arsenic in commercial vegetable shortenings. They are made from edible vegetable oils, by a process called "hydrogenation." An oil differs from a solid fat because an oil molecule contains less hydrogen than a molecule of solid fat. A process has been developed by which hydrogen can be combined with oil molecules and produce a solid fat. This is the process used in the manufacture of commercial vegetable shortenings.

Is honey more digestible than cane sugar? How do they compare in food value?

Mrs. J. R. B.

Both honey and cane sugar are completely digestible. Honey contains simple sugars, which require no further digestion. The process of digestion easily reduces cane sugar to simple sugars. Both are high energy foods, supplying approximately 100 calories per ounce.

What is the calorie value of apple butter? Of potato chips?

B. J. W.

Apple butter has approximately 45 calories per ounce, or two level tablespoonfuls. Potato chips give approximately 200 calories per ounce, or 20 large pieces. They are high in fat content, which accounts for the calorie value.

At what age can a healthy child safely discontinue the regular use of cod-liver oil?

Mrs. H. A.

[1]Dr. Carl P. Sherwin was director of the Good Housekeeping Bureau, which offered advice to *Good Housekeeping* readers on a wide variety of topics.

Carl P. Sherwin, "The Question-Box," *Good Housekeeping*, January 1944, 150.

Cod-liver oil is given to children because it is an excellent source of vitamins A and D. It is fairly easy to supply adequate amounts of vitamin A in the diet by including foods rich in this vitamin, such as deep-green, leafy and yellow vegetables, liver, eggs, milk, cream, cheese, butter, and vitamin-fortified margarine. Children require adequate vitamin D through childhood and adolescence, because it is essential for the proper utilization of calcium and phosphorus in building bones and teeth. In general, foods are not a good source of vitamin D. If the child gets plenty of sunshine and vitamin-D-enriched milk, it may be safe to stop giving the cod-liver oil. But you should consult your physician first.

Does bearing children cause the mother's teeth to decay?

Mrs. O. R.

Before accurate information on adequate diets was available, many women's teeth decayed or were lost after childbirth. The developing child needs calcium for bone building, and it must be supplied by the mother. When her diet does not furnish sufficient calcium, the embryo draws its supply from the mother's bones and teeth. An adequate calcium supply is necessary for people of all ages; but an extra amount is needed during pregnancy and lactation. Milk is the best source. A pregnant woman should have at least one quart of milk a day and a nursing mother one and one-half quarts, either as a beverage or in cooked foods. If a mother cannot take milk, the doctor will recommend calcium in another form. An adequate supply of calcium is essential if the baby is to have strong bones and teeth and if the mother's teeth are to stay in good condition.

A friend refuses to color margarine because she was told that butter-yellow rubbed on the skin of mice produced cancer. Is the color packed with margarine harmful?

Mrs. A. L. M.

Many persons confuse the term "butter-yellow" with the coloring provided for margarine. The color for margarine, after its purity is established by government analysis, is certified by the Food and Drug Administration. "Butter-yellow" is an old name given to another coloring. Applied to the skin of rats, it produced a cancerous condition. It is not the same substance as the natural color of butter or the color provided for margarine which is safe to use.

Do chocolate and cocoa contain oxalic acid? Is it harmful?
Mrs. H. B. G.

There is less than one-half of one percent of oxalic acid in chocolate products. It is entirely harmless. Oxalic acid, in small amounts, is found in many foods.

Please send me a list of foods containing carbon, hydrogen, and oxygen.
S. T.

All foods are organic compounds containing these three elements. The basic materials from which foods are formed are: carbon dioxide (a combination of carbon and oxygen) and water (a combination of hydrogen and oxygen). Plants, by the process of photosynthesis, use these materials to form carbohydrates (starches and sugars) and fats. Plants form proteins from these same materials plus nitrogen and sometimes other elements, such as sulfur.

Can the body assimilate the calcium in pasteurized milk?
Mrs. A. A.

Yes. Pasteurization does not affect the calcium content of milk or interfere with its utilization by the body.

Does adding vinegar to cooked vegetables destroy the vitamins?
J. G.

No. Adding vinegar to vegetables before serving has no effect on the vitamin content.

Figure 5. Star-Kist tuna advertisement from *American Home,* September 1950, 70.

29

M. F. K. FISHER[1]

"The Lively Art of Eating"

Harper's Bazaar, November 1944

Five thousand years ago Shen-nung wrote many wise pages about the irrefutable fact that we must eat to live and should therefore do it gracefully; and now in 1944 when I say that people who don't care much about their food don't care much about anything at all, I am, willy-nilly, quoting not only the Chinese king and Samuel Johnson[2] but several other philosophers, fat or thin but always jocund.

It is fortunate that such men are for the most part realistic as well. They think, which is their vocation, and once having thought, they write. They write about spiritual equations, but because they are philosophers and therefore realists even of the spirit, they write about the relations between hunger and love, between a dull palate and hatred or war . . . or inanition, worst of all. And I, Mosca the Gadfly, pip-squeak thinker, fetal craftsman . . . I can read what they have written and recognize it as the very truth. Yes, I can say wisely with Shen-nung and Plato and Johnson and Colette[3]—yes, since we must nourish our bodies in order to survive, let us turn this necessity from an obnoxious duty into something rich and gay and warm. Let us, I say with an almost ponderous nod, have fun.

And why not? Why can we not?

It is mostly the fault of our early training. In general the gastronomical hand-me-downs of Victorian England are heavy and stilted, and they are omnipresent in our literature as well as in our table manners. People ate a great deal then, but without conversation, and after their stupefying displays of opulence (plovers' eggs followed baked stuffed truffles, and were followed in proper turn by trout and salmon, game, roasts, ices, more truffles, omelettes, puddings, cheeses, heady pastes, cakes . . .)

---

[1]M. F. K. (Mary Frances Kennedy) Fisher (1908–92) wrote numerous books and essays on food.

[2]Samuel Johnson (1709–84), British essayist and lexicographer, was slovenly and ungainly in appearance.

[3]Colette (1873–1954), French novelist.

M. F. K. Fisher, "The Lively Art of Eating," *Harper's Bazaar*, November 1944, 88.

they retired to the drawing room because that was the way the Queen liked it, and looked sleepily at books of steel engravings.

In countries farther south, of course, such moralistic overtones were never heeded, and even a few Englishmen rebelled; but most of us were raised in families where it was still considered bad form to talk with any enthusiasm, if at all, about what was set before us. We were to eat it, be grateful for it, and then forget it as soon as possible.

Inevitably such an attitude of austerity bred a dour table, unworthy of remark. Good cooking, like a pretty woman, thrives on compliments, and as soon changes into something flat, shrewish, and discontented when it is ignored. The present state of English gastronomy, irrespective of the war, is proof of this. (And most probably the past four years have done more good than harm to people forced literally at gun-point to eat carrots, brown bread, and other such lusty foods!) Travelers from other lands have been shaking their heads with malicious sadness for several hundred years over the silent Britisher at his tasteless heavy table. "Good food should be talked about," they've said, "but John Bull treats it like an impropriety. He has countless religions, but only one sauce! And no wonder English bacon is so good . . . only the pigs eat well where all flavor is thrown away as something decadent and vulgar!"

In this country we have fared better, probably because of our mixed heritage, our grandfathers from Sweden and Bohemia and Sicily as well as from the bare downs of Sussex and Scotland. Even so we have been brought up, in great part, to feel that it is all right to gormandize on certain days of the year, and then regret it biliously for as many days after, but that for ordinary living "the plainest is the best." This is a direct hangover from Victoria,[4] and although it is basically true, its connotations are tangled in our minds with Being Good, with Behaving Ourselves, with Living Right. The sooner we divorce ourselves from such infantilism, the more fun life will be.

Lately it has become necessary, according to advertisers, to send questionnaires to men in jeeps and foxholes and gun turrets, asking them what kind of home life they want to come back to. Almost all the warriors whose letters are printable manage (among their many plugs for glass walls, thick plastic rugs, and invisible furnaces miraculously manufactured by the very companies who print their letters) to make it clear that as far as they are concerned the less kitchen there is, the better. They don't want their little wives tied to a hot stove the way Ma used to be. The kitchen, they say, must either be hidden in a couple of desk

---

[4]Queen Victoria (1819–1901) is associated with excessively repressive moral attitudes.

drawers or it must be built on the principle of the assembly line: the whole family enters at one end, moves with speed past a few electrical gadgets, and zip! everyone is out at the other end, fed with correct amounts of the latest vitamin concentrates to keep him going at top speed for several more hours. . . .

This, of course, is another proof of infantilism, or rather of the influence of women on men. Most females in America remember what a lot of work it used to be to get ready for some big indigestible family feast like Christmas . . . all the dirty dishes, the feverish fuss, the bad tempers. And they remember the deadly monotony of the rest of every year: roast on Sunday, hash on Tuesday, fish on Friday, with nobody caring enough to talk about it whether there were carrots-and-peas-in-white-sauce or green beans, as long as there were plenty of potatoes and pie.

Nobody caring . . . that was it! And the women cared least of all, because they and everyone they knew had been reared to think it woman's fate to bend over a hot stove, man's fate to bolt the results of hours of dreary thankless work and then hold it down with a digestive pill. There was revolt, finally, quiet, ruthless . . . and the result is that men are doggedly writing home from battle saying that *they* don't want their wives to waste themselves in dull kitchens, when of course it is the women who are saying it!

The other plans the men so blandly believe their own are equally interesting, in that they show the general taste of their mothers, wives, girl friends as dictated by the "home economists." They don't want dining rooms, most of them: they plan to eat as simply and rapidly as possible, and want to "go out" quite often, so why give Honey-lamb another room to take care of? And they don't see the point of cellars. After all, heating and so on will be done with buttons (and a few mirrors!), so what would anyone do with a cellar? Honey-lamb couldn't put up peaches if she had to . . . and the canned ones are all right, although not of course as good as they used to taste from Grandma's preserves shelf . . . and what else would a cellar be for except for rows of glass fruit jars? (Of course before the war you'd have made a game room of it, but in the new plans the living room is big enough for that too, with ping-pong tables of plastic, a television screen over the electric fire log, and a bar that mixes a Martini, gives the latest stock reports, and plays music to any room in the house, so that the children can be sung to sleep upstairs while their parents are lulled in another manner down under the infra Z lighting. . . .)

It is too easy to make gags about (and to gag over) the so-called Postwar Dream House, with all it may mean in our whole pattern of living. One thing that can never be even faintly funny, though, is that men who have been living on dehydrated food, pulled any-which-way from little ration tins, not only may not be able to taste good food, but may not be able for a time at least to stomach it. Their poor tight guts will growl and repulse as poison the green salads and the rich fine sauces they have sighed for and once welcomed.

They will, perforce, look at every meal with a hungry but dyspeptic eye. They will dream sadly of the way they used to eat flannel cakes when they were kids . . . and will go wearily out to one more meal with Honey-lamb at The Greene Arbor Tea Roome, where there is "home cooking by women chefs" and a choice of twelve relishes on a tray, free on the $1.25 dinner or 10 cents extra on the $1.10. Nothing will be said about the food while the two people eat, or before or during or after the boring meal, because it is obvious that there is nothing much that can be said.

This picture is a grim one, especially when it is stripped to the factual bones: men and women spend most of their time together either eating or in bed. They are hungry. They must eat in order to do a lot of things such as making love. They can't do one well unless the other matches it, and feeds it, and makes it worth doing.

When the men come home from war, they should know the sweet comforts of the flesh, so that their spirits may feel strong. They should see the reassurance of shelves glowing dimly, like stained glass, with the fruits of harvest put away for the future, for them and their children, in cellars or cupboards. They should see bottles of good honest wines, too, lying on their sides waiting, and the promise of beer cooling in the darkness.

They should feel in their nostrils the sting of memory: a stew simmers in a clay casserole at the back of the stove, even if it is a two-burner arrangement in a walk-up flat, and because of a whiff of rosemary it is like the chicken they ate once in a little red-ink joint outside Dallas, or because of the sherry added at the last it reminds them of a kid brother's christening party, when everyone was so young and happy. . . .

And they should find Honey-lamb at ease, not hot and flustered and resentful, not filled with a vision of the last month's "recipe page" in her favorite slick magazine and therefore tangled hideously in a web of fancy sauces and uncomprehended flavors, and as edgy as a colt in May.

Honey-lamb must be lost in a sensual realization of all the magic she can evoke at table. She must be full of enjoyment, of a rich savoring of the odors, the textures, the colors of a good meal. And she must be full of patience most of all, with her nerves lulled, so that there cannot possibly exist a puckered brow, a cold turning of the head with tiredness or ennui. She must have the belief about her that preparing food, together or alone, is an act of dignity, and that sharing it with someone she loves is akin to love itself, so that all the other good things follow naturally.

Then there will be a loosening of the snarls, and a smoothing of the rumpled lonely thoughts, and soon the bad happenings like doubt and ack-ack and tight bowels with the man, and solitude with the girl, will be wiped out by the shared peace of meals together.

At first there will be one simple dish, or two. Gradually ease will flower, and then the immediate past and the further one of strained "family feasts" will vanish, and every plate will be honest, and something that can be talked about.

Talking is important. Good food can be discussed without gluttony. It can be pondered without piggish gormandizing. When the children come it will be as natural a part of their lives to talk out loud of the salad or the little round flat ginger cakes as it was of their parents' to remember them secretly as something fleshly and therefore shameful.

And best of all, the father and mother and all their thriving babes will have a part in the feeding of the whole, so that what used to be a dogged female duty, a dyspeptic necessity celebrated somewhat orgiastically at Christmas and Thanksgiving, will now become a family pleasure, something shared daily with mutual enjoyment. Men no longer will be shoved from the kitchen as sissies or annoyances. They'll need no longer boast pathetically of the one fabulous, histrionic sauce or salad dressing they are permitted to make while their polite impatient wives look on humorously, and secretly tot up the pile of extra bowls and dishes to be washed. And the children will never stifle the passionate love they may feel tonight for a custard, tomorrow equally for a slice of cold tomato with one leaf of basil on it. Meals will be fun.

They will be fun, yes, in spite of early training, in spite of postwar planning to eradicate them. Kitchens will have a thousand gadgets, naturally, that sound miraculous now and much too perfect to be true. But if the kitchen is made part of life, and not a thing to be shoved into a closet like a shabby skeleton, it will keep more wolves than the wolf of hunger from the door. It will push back the frontiers of boredom . . . and they are, probably, the farthest and the bleakest.

The warm full lively kitchen will make tired men calmer, and little children gayer, and it will make women everywhere live more magnificently than if they have only to push a few illuminated buttons here and there and with a swish and a yawn feed their broods, thanking God as they do so that things have changed since Grandma was a bride.

By the time women have felt their own power in their own kingdoms, they will know that they have never been slaves, except to a mistaken idea . . . and that eating to live, with the herbs and the skillet and the shared flavors of existence can be as rich a part of life as ever living to eat has been a poor one. They will know that no matter how banal this sentiment may sound, it is still quoted from the best philosophers, and therefore safe.

30

DOROTHY THOMPSON[1]

"Occupation — Housewife"

Ladies' Home Journal, March 1949

A woman of my acquaintance, whom I interrupted while she was filling out an official questionnaire, laid down her pen with a sigh. "One question on all official or unofficial papers—government, legal, tax, what not—always gets my goat," she said. "Fills me with an inferiority complex as 'deep as a well and as wide as a barn door.' It's that query 'Occupation?' And I have to write down 'Housewife.' When I write it I realize that here I am, a middle-aged woman, with a university education, and I've never made *anything* out of my life. I'm just a housewife."

I couldn't help bursting into laughter. "The trouble with you," I said, "is that you have to find one word to cover a dozen occupations all of which you follow expertly and all more or less simultaneously. You might write: 'Business manager, cook, nurse, chauffeur, dressmaker, interior decorator, accountant, caterer, teacher, private secretary'—or just put down 'philanthropist.' "

"Philanthropists are people who give away money," she demurred.

"Not in the exact meaning of the word," I countered. "A philanthropist is a person who loves humanity and gives away something for love. All your life you have been giving away your energies, your skills, your talents, your services—for love."

---

[1]Dorothy Thompson (1894–1961) wrote a regular column on political and social issues for *Ladies' Home Journal* from 1937 to 1961. After working as a foreign correspondent for the *Philadelphia Public Ledger* and the *New York Evening Post* in the 1920s, Thompson wrote a column titled "On the Record" for the *New York Herald Tribune* from 1936 to 1941. Her books include *The New Russia* (New York: Henry Holt, 1928), *Dorothy Thompson's Political Guide* (New York: Stackpole, 1938), and a compilation of her newspaper columns, *Let the Record Speak* (Boston: Houghton Mifflin, 1939).

Dorothy Thompson, "Occupation—Housewife," *Ladies' Home Journal,* March 1949, 11.

"Not exactly giving them away," she doubted. "I've been supported. And I guess I've been paid with lots of love in return." She looked happier.

"Or you might put down 'Free woman.'"

"Free?" she countered. "I can hardly call my soul my own."

"Oh, yes you can," I replied firmly, for the woman is a great friend and I know her very well. "And your life has always fulfilled the prime test of a free existence—namely, never to do anything just for money."

"Perhaps you have something there," she smiled. "But here I am, nearly fifty years old, and I have never done what I hoped to do in my youth. Music! I played the piano twenty-five years ago better than I do now. And I had a college education—wasted."

Wasted! Without a mind trained to concentration, to tackling and solving problems, weighing alternative policies, and planning the use of time, this woman never could have done what she has. As for her artistic and intellectual interests—

"But all your children are musical. And that is simply because you brought music into your home—isn't it?"

My friend grinned. "But all this vicarious living—through others"— and she sighed again.

"As vicarious as Napoleon Bonaparte," I scoffed. "Or a queen. I simply refuse to share your self-pity. You are one of the most successful women I know."

And I certainly meant it.

This woman married, at twenty-one, a struggling high-school teacher. They went to housekeeping on $35 a week. They had three children— two sons and a daughter. It took her husband twenty-five years to rise high in his profession, and during fifteen of those years—until an outstanding and popular historical work brought him to the forefront—they had to live on very modest means with exceptionally refined tastes. Their children were grown before teaching and books won him a quite handsome income. But if this break had never happened, my friend would have been no less a success, for her greatest achievements were performed when her husband was earning between $3000 and $5000 a year. During all that time her husband and family never lived in anything but a well-kept, charming home. They never in their lives ate a bad meal. They were always attractively dressed. All three children were (and are) admired for their good manners and exceptional intelligence and industry—which saved their parents much money, for they were largely educated by scholarships. They were also educated for kindness and consideration, because there was always room for one more in that home, and for several years it was shared with a refugee child.

To do what this woman did with her husband's modest income was a feat of management, showing executive ability of a high order. Her gifts as a craftsman were no less formidable. I once found her papering her living room—"a perfect cinch once you get the hang of it," she remarked happily. As a food buyer she would have won a high salary in any restaurant, for she watched the markets like a hawk, and planned delicious meals accordingly. When the children were small she made all their clothes and most of her own, including suits and topcoats, having gone between housekeeping duties to a tailoring school. In "time off" she typed her husband's manuscripts and proofread every book; played piano duets with the children to make practicing more fun; followed them in their reading when they were in high school, the better to discuss the books under study; and as they grew up and went away to college, threw herself into the work of the community: sat on city housing boards; planned festivals to make up the church deficit; took the lead for better schools; and, in fact, when I think of her solid achievements they are matched by few "career" women.

"But I never earned any money." That is the lament of many "housewives."

If the family is considered as a unit, that is simply not so. Millions of women, all over the United States, are contributing as much to the well-being of their families by the services they render and the brains they mobilize, as are their income-earning husbands. Compile the cost of their services, if it had to be paid to half a dozen professionals! And what do most men work for anyhow? They work for their homes! The home they are able to support is the real measure of their financial "success." In that home expenditure is quite as important as income, and 80 per cent of American income is spent by "housewives," for better or for worse. Who does not know $10,000-income homes that cannot match in order, comfort and comeliness homes run on half that income? Who supplies the economic ability which overcomes income deficiencies? Invariably some "housewife."

Who can hire, at any price, a substitute for a mother? Who can find a housekeeper who thinks twice about every purchase, weighing value against available cash? On what labor market is affection to be purchased, patient devotion, good humor, laughter? There is no question that most women can save in the home by their managerial talents more money than they could bring into it from outside work.

How many men would have given up in despair in those troughs of life which come to everyone, but for the patient faith and the carefully concealed sacrifices of a loving wife? That woman of despised occupation!

It seems to me I am continually opening novels about the pretty girl who turns into a slut and a harridan upon marriage; or whose spirit is broken by the boredom of household tasks. Perhaps I know only very nice people, but I know them in all walks of life, and it seems to me that these dismal creatures are the exception, not the rule. It is probably true, as Mrs. Virginia Woolf pointed out in *A Room of One's Own*,[2] that lack of money of their own, and of leisure, has thwarted the genius of some women. It is true that most conspicuous women geniuses have been childless, and free from household cares. But the world and civilization are not made by genius alone, but by civilization's myriad unsung heroes and heroines. And a world full of feminine genius but poor in children would rapidly come to an end.

One might also ask: What would become of the men geniuses? Someone has said "When you see a great man, you can induce a great mother." That has been confirmed over and over again, and by great men themselves. But great mothers, like great geniuses, have to work at their task. It isn't just an inborn talent that flourishes without constant effort, and in free time. And most good men had *good* mothers too. Children—especially boys—usually get their ethical standards, as well as their ambition and courage, largely from their mothers.

The instinct of the masses of the people in Catholic countries, who without benefit of theologians elevated the mother of Christ into part of the Godhead—as eloquently told by Henry Adams in *Mont-Saint-Michel and Chartres*[3]—was a sound one, growing out of human experience. "If Christ is perfect," they argued wisely, "His mother is perfect, for only a perfect mother has a perfect son." And so they elevated the gentle and inconspicuous Mary into the Queen of Heaven, and built their loveliest cathedrals to "Notre Dame"—Our Lady.

That may seem a far cry from "Occupation: Housewife." But is it so far? The homemaker, the nurturer, the creator of childhood's environment is the constant recreator of culture, civilization and virtue. Therefore, assuming that she has done and is doing well that great managerial task and creative activity, let her write her occupation proudly: "Housewife!"

[2]Virginia Woolf (1882–1941), English novelist and essayist, wrote *A Room of One's Own* in 1929.
[3]Henry Adams (1838–1918), American historian and writer. His philosophical work *Mont-Saint-Michel and Chartres* (1913) discusses the Virgin of Chartres as the link between science and religion.

31

HELEN W. KENDALL

"Electric Mixers — Strong Right Arms"

Good Housekeeping, January 1950

In every kitchen at Good Housekeeping there is a strong right arm—and it isn't the cook's. It's an electric mixer that takes the work out of cooking. We're so accustomed to letting mixers do the work, we wonder why a woman tries to do without one. When we talk to women in homes, we find that some use their mixers "for everything"; others leave them idle except when they have to do a heavy mixing job—such as mashing potatoes. Here are their reasons:

1. *"It's too much trouble to set up the mixer."* Inevitably, when a housekeeper makes this comment, we find the mixer tucked away instead of on a table or counter where she does her mixing. Does this apply to you? We urge women to put their mixers in a convenient place and set them up with bowl and beaters, near an electrical outlet. If you can leave it plugged in all the time, so much the better.

2. *"It's easier to mix by hand."* Women who say this haven't realized the mixer's possibilities for saving time and work. A mixer isn't just a cakemaker or an orange juicer; it's a tireless arm for pesky jobs, such as creaming butter and beating frosting, and a speed demon for making meringues, beating eggs, and whipping cream.

3. *"I make only package-mix cakes now."* That's all right, but what's the matter with using the mixer for these and for quick-method cakes of any kind? Three hundred strokes of beating by hand can seem like an eternity, and a mixer does it so much better. Maybe you have hesitated to use the mixer for fear of overbeating the cake at the "low" or "medium" speed, whichever the directions call for.

While mixer manufacturers have gone a long way in numbering speed dials and designating the right speed for most jobs, as yet none of them has a position for quick-method cakes or packaged cake mixes. So until we made more than 100 cakes with all the leading mixers, we were as

Helen W. Kendall, "Electric Mixers—Strong Right Arms," *Good Housekeeping*, January 1950, 130.

uncertain as any housekeeper would be about what "low" or "medium" meant when mixer dials have numbers for speeds.

Packaged mixes: When the directions on the box called for "low" speed, we made the best cakes at one or two settings above the lowest speed. When they called for "medium," our cakes had the best volume and texture mixed at a spot in the center of the dial—for example, No. 5 for a 10-speed mixer, No. 8 for a 16-speed one.

Quick-method cake recipes are those in which dry ingredients, shortening, and some liquid are beaten together first, with the rest of the milk and eggs added and beaten together later. In making these cakes, we tried several different settings above, below, and at the center of the dial (medium). All were satisfactory, but in one or two cases, our judges felt that the peak of perfection was reached at the speed just below "medium." Try making cakes of this kind with your mixer set at both "medium" and slightly below, and do a little judging on your own. If recipe specifies "low," see "Packaged mixes," above.

32

JANE WHITBREAD[1] AND VIVIAN CADDEN

"Granny's on the Pan"

Redbook, November 1951

Men who dote on Mamma's cooking are living in a dream world. That's the scrappy verdict of two modern young wives who throw everything but the kitchen sink at tradition.

The Mamma's-apple-pie fixation of most husbands is one of the deep-seated neuroses of our time. It is based on the dubious premise that only men had mothers, and only men's mothers cooked. At the risk of spoil-

---

[1]Jane Whitbread, who also wrote as Jane Whitbread Levin, worked as an advertising copywriter for the film industry and began her career as a freelance writer in 1937. Her works include *The Intelligent Man's Guide to Women* (New York: Schuman, 1951), *How to Help Your Child Get the Most out of School* (Garden City, NY: Doubleday, 1974), and *Daughters: From Infancy to Independence* (Garden City, NY: Doubleday, 1978).

Jane Whitbread and Vivian Cadden, "Granny's on the Pan," *Redbook*, November 1951, 44.

ing the story, we would like to point out that we had mothers, too. Besides, having eaten a bite or two of Mamma's fare, we have a few memories ourselves.

By dint of Mamma's cooking a child began his life with colic and ended it with obesity, the whole span being liberally spiced with bilious attacks, heartburn, acute indigestion and a touch of ptomaine poisoning. Not that Mamma wasn't a good cook.

There was her oatmeal, for example. By the time it had cooked in an iron pot for eight hours it was almost as smooth as laundry starch. The last one down for breakfast was lucky enough to get the crisp burned part and the job of setting the pot to soak until it was time to start the cabbage on its four-hour simmer for supper.

If you didn't like oatmeal, there were always Mother's famous hot breads. The hot breads were pretty good if the oven temperature was steady that day. Even when they weren't up to scratch, they at least stuck to your ribs by sheer weight.

Mother certainly had a way with meat, and that way was gravy. We can't seem to remember whether it was beef or veal or lamb. It really didn't matter, because the end product was always the same. From this comes the expression "stewing in your own juice."

Nor can we overlook Mamma's chefs-d'oeuvre—the apple pie and the homemade ice cream.

We have a theory that the apple pie that every husband remembers was the one Mamma made one day in August. The apples had ripened but not yet browned, and the worms had had a worse season than usual. This was the day when Mamma had chance on her side. For once her casual proportioning of flour, salt, shortening and water turned out a confection you could hardly tell from a modern mix.

Besides, that was the day Sonny got his first bicycle. Anything he ate would have been memorable.

Aside from an occasional piece of rock salt, Mamma's ice cream was undeniably good. The only reason it isn't on the tip of every man's tongue is that no one ever got enough of it to remember it very well. It was such a large-scale operation that Mamma never thought of undertaking it unless she was guaranteed a minimum eating audience of two dozen. Since the freezer usually held two quarts, everyone got at least a taste.

Mamma's table certainly had eye appeal. It was a colorful diet she served up. Potatoes in off-white provided a stunning contrast to vegetables in olive drab or putty beige.

It was hard to tell whether green peas or beans or spinach were a-boil on any particular day. The hereditary smell of cabbage, the trademark of every home, obliterated everything weaker than boiling cod.

The least you can say for Mamma is that she made the most of what she had. But let's face it; she didn't have much. During the winter she was handicapped by a supply of vegetables that owed their precarious existence to last summer's weather, the temperature of her cellar and the length of time it took for a cake of ice to become a river on the kitchen floor. Most vegetables were stored underground, grew underground, and never should have been dug up in the first place. Mamma's cooking equipment consisted of a series of cups and spoons of indeterminate size, of pots with a tendency to burn and lids that didn't fit, and a stove that could easily heat up the kitchen without bringing water to a boil.

The fact is that the whole food business has been so violently revolutionized in the past thirty-five years that today's pot roast would hardly recognize Mamma's as a blood relative. It took grade-A ingenuity as well as spice and gravy to turn the stringy beef of pre-4H-Club days into anything a saber-toothed tiger would eat. As for yesterday's bean—it was a bean. Today's is crossbred, bone-meal-fed, DDT-sprayed and French-cut. And today's peach, strawberry, raspberry or broccoli frozen at flavor peak would shrink at the thought of the fruits and vegetables most mothers squirreled up to brighten the turnip months. Without being catty, it took almost as much loyal devotion on the part of the family to empty Mamma's preserve shelf as it took her to fill it.

The bride of today may not know kohlrabi from rutabaga, but she can buy all the artichokes she wants with cooking directions and a recipe for hollandaise sauce printed on the wrapper. She may not know a fast boil from a simmer, but the odds are that her stove is plainly labeled to let her in on the secret. If a recipe says "chill for three hours" she doesn't have to bite her nails for fear the ice will melt before the time is up.

Part of the nostalgia surrounding Mamma's cooking derives from the simple fact that it took so long. The line about "sweating all day over a hot stove" was no joke in her day. When Dad came home he didn't need to ask, "What have you been doing today?" What she had been doing was all over the table, and no one could fail to notice it regardless of how it tasted.

The net effect of modern advance in the raising, marketing, storing, processing and preparation of food is to reduce the role of Mamma to a minimum. Today's Mamma probably spent her afternoon at a PTA meeting and started her apple pie at 5:30. By 5:33 she had rehydrated the apples, rolled the crust and swept it into a 450° oven, while smoking a cigarette, removing the week's wash from the dryer and getting next morning's breakfast menu from the radio. When she hears her husband's footsteps on the walk she throws the lamb chops under the broiler, the potatoes into the pressure cooker and the frozen limas into the saucepan. Before he has a chance to start sniffing, the dinner is ready. It is bound to be good—and it's not her fault.

For modern methods have made a science out of what used to be a lifelong experiment based on instinct, intuition and pure chance. Nowadays a woman can get a whole meal without knowing anything more about cooking than how to boil water. *And the result will be perfectly adequate.* On the other hand, if she's a first-rate good cook and likes to venture into the culinary unknown, her path is well charted by Fanny Farmer, "Casserole Cookery," Escoffier[2] in English, not to mention half a dozen monthly magazines such as *McCall's,* whose recipes are kitchen-tested for accuracy, vitamin content and man appeal.

Just because cooking is no longer a full-time job for a housewife, there's a tendency to overlook the results. Who can praise anyone who takes less than half a day to prepare for a dinner party, even if the dinner is a success?

Mamma's boy is apt to conclude that Mamma was a creative artist while his wife just follows recipes. And he's partly right. If Mamma ran out of shortening and somehow devised some baked goods without it, the result became a tribute to her ingenuity and everyone took their baking soda in private. The recipe inevitably became part of the family folklore.

It's high time that America's pressure-cooking wives stopped being apologetic because their family life doesn't revolve around the kitchen. Explosion of the myth that Mamma liked nothing better than to spend all day trying to whip food into a semiedible state is long overdue. If she were still alive and cooking, she would be the first to latch on to preseasoned, quick-frozen hamburgers and French fries topped off with pre-rolled pie crust loaded with canned blueberries. If men would just skip this malarky about 19th-Century menus and eat what's put before them, they'd be better off. Unless a man marries a woman to get rid of her, you'd think he'd rather have her spend more time with him and less in the kitchen—especially when he can have his cake, too.

The truth of it is that Sonny's romantic memory of Mamma's cooking is just a classic example of the Oedipus complex at work. If Freud is still

---

[2]Fannie Farmer (1857–1915) began Miss Farmer's School of Cookery (1902) in Boston to train housewives in cooking. She wrote many instructional books on cooking, among them *The Boston Cooking School Cook Book* (1896). Georges-Auguste Escoffier (1846–1935) was a French chef who also taught and wrote books about cooking.

in vogue thirty years from now, we predict that husbands will look up from their meal of vitamin tablets and say with that old wistful expression, "When I was a kid Mamma made 'brown-and-serve' rolls every day."

33

ROBERT J. KNOWLTON[1]

"Your Wife Has an Easy Racket!"

American Magazine, November 1951

If the little woman has a quick temper don't let her see this round-the-clock picture of her day. It's based on the author's scientific survey of housewives' activities. In fact, better not let her read this article at all, since it tells how she can have even more of a cinch.

I know I am leading with my chin when I say this, but the fact is that, contrary to what your wife has most likely been telling you for years, she has an easy job.

Actually, she works about 6 hours a day as a general rule and almost never more than 8, unless in extreme emergency. Frequently, her workday is no more than 4 hours. The amount of time varies with the size of the house and the family. Furthermore, she would not even have to work as long as she does if she did her job well. If she wasn't stubborn, tied down by tradition, a poor organizer, and even just plain lazy, she could be free most of the day.

She spends hours looking at television or listening to radio soap operas, but doesn't deduct this time from her workday. She entertains or visits neighbors, and calls this part of housework, when all she's doing is gossiping. She includes as labor even the naps she takes during the day. She reads books and magazines and still says her job is killing her.

As I said, I'm leading with my chin, but these conclusions are not

[1]Robert J. Knowlton was a professor of industrial engineering at Northeastern University.

Robert J. Knowlton, "Your Wife Has an Easy Racket!" *American Magazine*, November 1951, 24.

mine. All I did was conduct a coast-to-coast scientific survey and these results came to light. More than a thousand families participated. I made a round-the-clock study of my own wife's activities for a period of weeks. My students in industrial engineering did the same, using their wives, mothers, and married sisters as guinea pigs. I got detailed reports from colleagues of mine, former students, and friends scattered across the country, whom I asked to conduct the same sort of research in their own homes. In addition, these investigators made studies among their friends and neighbors, enlisting the men in the families to act as "spies." All sent in their findings.

Some lived in cities, others in suburbs and small towns. Some of the women had from one child to 4 children, a few had none. Ages of the children ranged from infants to "grownups" in their late teens. This gave variety to the survey.

For a week we checked the women openly and aboveboard, and in this way we found out about their daily schedules. However, we discovered far more later, when the men began watching their wives surreptitiously.

Perhaps I might best begin by explaining how I came to launch this project. I learned the hard way how easy it was to run a house. For 7 months, while my wife was having a rather difficult time bearing our daughter, Amanda, I was the housekeeper of our 4-room Cape Cod cottage. I also had a full-time job teaching.

At home I cooked all the meals, made the beds, cleaned the house, and even took care of the cat. For the first few weeks it was pretty rough going. I always seemed to be in a rush and never got things done. If anyone said a housewife had a tough time of it I was the first to agree. I couldn't wait for our baby to be born so I could retire.

Then, one evening, a girl-friend of my wife's dropped in to see us, and before leaving she inspected the house to see how I was faring. In the kitchen she found dirty dishes piled up in the sink. Half-empty boxes of food were strewn about. Crumbs crackled under your feet as you walked across the linoleum. I sheepishly told her she had caught me just as I was going to work on the kitchen.

But in the living-room dust was mounting on the furniture, and all sorts of odds and ends were standing in corners waiting to be placed in the cellar, the attic, or a closet. I told her I was just about to clean up the living-room. However, when it was no different in the bedrooms I gave up excuses and said nothing.

"Nice mess you have here," she remarked, taking off her jacket. "Come on; let's clean it up. I'll help you."

Then she dropped a barb that hit me right where I lived and started my reformation.

"You know," she said, "you're typical of the do-as-I-say-don't-do-as-I-do school of thought. As an industrial engineer, all day long you're teaching people how to plan and make work easier in industry. Then you come home to your job here and create complete chaos, by violating every principle you preach."

That did it. You girls may say that it took a woman to show me the way, and I'm delighted to give the credit where it belongs. The important thing is that I began applying science to housekeeping and as a result learned what an easy job it really is.

First of all, I arranged all cooking utensils, dishes, and silverware in drawers and on shelves nearest the spots where they would be most used. I did the same in storing food in bins, closets, and refrigerator. This took a little experimenting, but when I had everything set I found the arrangement saved loads of time and effort.

I analyzed the problem of dishes and found that the reason I let them pile up was because I detested drying them. This was a holdover from my teen-age days, when I had to dry them every night. I solved this simply by not wiping them. I scalded the dishes as I washed them, then let them stand in the draining rack for a few minutes while I tidied up the rest of the kitchen. By that time they were dry and I put them away.

As for taking care of the other rooms, I discovered the job looked big because I hadn't kept up with the clutter. I had let things go. But now, when I saw dust I cleaned it off immediately. If a rug needed cleaning I cleaned it. I put things away after using them, instead of letting them accumulate. This new way, I was always doing little jobs instead of big, exhausting ones.

In this connection, I also learned that the old-fashioned idea that certain days had to be set aside for certain jobs was a waste of time and energy. You probably recall the schedule: Monday, wash; Tuesday, iron; Wednesday, sew and mend; Thursday, day off; Friday, bake; Saturday, clean; Sunday, church and cook. I found it much easier to do all these jobs as I went along. I would wash and iron three times a week, dust and pick up every day, and so on.

I discovered I was getting the job done in no time, and wondered why housewives were always complaining that "a woman's work is never done." I wondered, too, how much work a housewife actually did, and how she managed to make it look so hard. That's when I got the idea of conducting the survey. I might say in passing that I followed my own wife around with a stop watch and wrote a report. The other investigators did the same, and when I put all these reports together, the conclusions varied hardly at all. All the men found conclusive evidence that their wives had pretty much of a cinch.

They also made some interesting and enlightening discoveries.

For instance, they suddenly realized that previously they had had no idea of what their wives had been doing during the day—except that it must have been hard work, because their wives told them so. You know the familiar line—"While you're in your air-conditioned office ogling your beautiful secretary, I'm here at home working my fingers to the bone over a hot stove."

It has not been too hard for women to convince their husbands of this, because many husbands do "ogle" and, besides, many of them recall that housework was real drudgery when they were growing up—in the days before the extremely wide use of automatic laundries, dish-washing machines, vacuum cleaners, pressure cookers, and many other labor-saving devices. Most husbands also grew up in an era of bigger houses, which were more ornate and more heavily furnished than today.

At any rate, when the men began clocking their wives and doing some sharp house detective work, they discovered some illuminating facts about which their wives hadn't bothered to tell them.

For instance, one of the husbands reported that on a certain morning at breakfast, his wife seemed to be bustling with plans for the day. She had a list of projects and, the way she told it, she'd be lucky if everything got done by midnight.

He kissed her good-by. A few minutes later, while driving to town, he had the inspiration to see her in action when she didn't expect him, so he called his office to say he wasn't coming in that day. Doubling back, he arrived home about 10 A.M. He found his 6-year-old playing happily in the front yard, but he heard no sounds of whirring activity within the house.

"Where's Mother?" he asked Junior.

"In bed," was the reply.

"Is she ill?" he asked, at first alarmed, although she had been positively glowing with health when he left.

"Naw," squealed Junior, "she goes back to bed every day after you leave."

Now, this husband had no objection to his wife's going back to catch a few winks, but the next time she moaned about how hard she was working, he wasn't too impressed. He was more inclined to inquire, "What happened? No nap today?"

Another "Household Hawkshaw" had his eyes opened when he arrived home too early one afternoon at about 3:30 P.M. instead of his customary

6 P.M. He found his wife sitting in the kitchen chatting happily over a cup of tea with a neighboring "slave." She was still wearing the housecoat she had on when he had left in the morning. The dishes had not been washed, the beds were unmade. He discovered this was a usual state of affairs.

He soon learned the truth—that his wife had been doing all the housework in a mad rush just before he was due to arrive from work. No wonder she had been puffing and tired when he walked in the door.

Practically all the scientists who participated in the survey disclosed that during the midafternoon it was a usual thing for the mothers to gather with their children to play and chat in small groups in front of houses, in a neighbor's garden, or in a park. This was practically a daily ritual, especially if the weather was fair.

Then, about 4:30 P.M. a strange thing would happen. There would be quick good-bys and "see-you-tomorrows," and almost in a flash the streets would be deserted, the parks emptied. Inside the homes there would be sounds of great activity, while the mothers, in a frantic rush, did the day's housework and prepared dinner. Small wonder that when the husband came home he found his wife perspiring and complaining of what a day she'd had.

Another observer, a bachelor, ran into a situation which, if he'd been inclined to "sing," would have caused a major domestic upheaval. Having no mate of his own, this lad checked upon his married sister.

Now, Sis had 3 children—all small—and her husband was a traveling man who came home only on week ends. Although this innocent ram provided his lamb with all the modern conveniences, he was drafted week ends to help with the heavy cleaning and shopping. Sis told her everloving that she was so bogged down during the week with the children and other household duties, she "just couldn't do everything."

However, her nosy little brother gleaned the information that Sis spent several hours each weekday entertaining neighbors, looking at television, reading, and napping. All good, clean fun, but she called it work. Brother filed his report and marked it CONFIDENTIAL.

All my "spies" made detailed reports on hour-by-hour activities of their subjects and they ran pretty much the same. Since it is typical, I will give you a summary of my wife's schedule at the time I made the survey, after Amanda was born. It went like this:

7:30 to 8:15 A.M.: First three quarters of an hour was spent attending to normal needs, taking care of the baby, preparing and eating breakfast.

8:15 to 9 A.M.: Cleaning up kitchen, stacking dishes, feeding cat, doing the wash in basement, taking baby with her. (If we had owned a washing *machine, which we didn't, the time in the basement would have been shorter and she could have done other things while the clothes were in the machine.)*

9 to 9:30 A.M.: Playing with baby, taking a breather. (This 30 minutes not counted as actual work.)

9:30 to 10 A.M.: Hanging clothes and watering plants.

10 to 10:30 A.M.: Making beds and cleaning up bedroom.

10:30 to 11 A.M.: Carpet and rug cleaning, or cleaning out a closet, or basement; varnishing or painting.

11 to 11:15 A.M.: Dusting and cleaning living-room.

11:15 to 11:30 A.M.: Preparing lunch for baby and self.

11:30 to 11:45 A.M.: Eating lunch (not regarded as work).

11:45 a.m. to 3 P.M.: Planning dinner and making out grocery list, while baby naps. Then relaxing with a book or listening to radio (not regarded as work).

3 to 5:30 P.M.: Out with baby in the carriage. (One hour of this time is spent in shopping and can be called work; the rest constitutes gossiping and enjoying the fresh air.)

5:30 to 6 P.M.: Preparing dinner.

6 to 6:30 P.M.: Eating dinner (not regarded as work).

6:30 to 7 P.M.: Dishes, cleaning up kitchen, giving baby her bath and putting her to bed.

7 P.M. to bedtime: Relaxation—no more work.

If you'll add up the hours of the day shown here and subtract the free time, you'll see that it comes under 6 hours. Of course, this schedule varies slightly from day to day, but the total time does not change much. On the days my wife doesn't wash—she does so 3 times a week—she irons. When she has no clothes to hang, she spends the time baking cakes or cookies, or mending. I observed that the amount of time she had for leisure changed hardly at all from day to day.

Now, I can hear some mothers screaming, "Wait till that kid gets older and see what problems you have."

In the first place, if the child is properly trained, he or she will not be too much of a problem. Besides, I do not regard as work the training of children, which should be shared by both parents. That's supposed to be the real pleasure of married life. Finally, as the children grow older, they spend several hours a day in school during most of the year, leaving the mother free for other pursuits.

There was little significant difference in any of the schedules reported by my observers. For women with larger families and larger houses, the day's work was closer to 8 hours. Furthermore, families with most of the modern conveniences disclosed that the housewife worked considerably less than 6 or 8 hours.

Probably the most convincing proof that the housewife has a lot of free time was the fact that so many reports showed women engaged in outside activities, for either pleasure or profit. One woman sold cosmetics and another did part-time nursing. One mother gave driving lessons and another reported the neighborhood news for a local paper. Many of the women spent several afternoons playing golf or tennis, taking groups of youngsters coasting in winter and on nature hikes in summer.

If any men reading this think I'm exaggerating all you need to do is a bit of quiet checking on the little woman when she thinks you're reading the paper. Start clocking her.

You know, the funny part of all this is that after our survey was completed and the results were added up, I got to thinking that, easy as a housewife's job is now, it could be made even more of a cinch if she would apply a few of the short cuts of industrial engineering. Should I let her in on the secrets? Why not? So here are 10 simple suggestions:

1. *Ironing:* Any woman can cut the time in half by eliminating needless ironing of things nobody ever sees, such as underwear, pajamas, sheets, pillowcases (if covered by a spread), handkerchiefs, and play suits. I included play suits because, while they are seen, they get mussed so quickly it's hardly worth the bother of ironing.

2. *House cleaning:* About twice a year most housewives tear up a house, cleaning it from top to bottom. This is unnecessary. Why not clean each part of the house as it needs it and forget the big cleanup! That's the way it is done in a well-operated industrial plant.

3. *Cleaning tools:* Have several dust-cloths, pans, mops, etc., on different floors or in various parts of the house. Many jobs pile up in a house because the mop or cloth is not handy at the time you see a part of the floor or wall that needs cleaning.

4. *Washing clothes:* There are many short cuts here. In removing clothes from washing machine or tub, place them in the clothes basket in the order they will be hung. If you are building a house, make sure the basement door is wide enough for a woman to get in and out with a basket. Use a rolling table to carry basket to clothesline (a baby carriage can also serve). Work from the farthest point back toward the house and always hang the biggest and heaviest stuff first. Vary the height of the line. Stretching is very tiring. You need a high line only for the big pieces, such as sheets. For smaller pieces have the line about chest-high. Of course, the ideal thing would be to have an automatic drier.

5. *Kitchen and dining-room:* Have table on wheels to go from dining-room to kitchen to pick up dishes and food; also, use lots of large trays and save trips. Many women like to keep all their dishes together, especially if they have a complete set. However, considerable time and energy can be saved by locating table dishes in the dining-room and serving dishes in the kitchen. Do the same with silverware. Arrange kitchen items so that they are closest to place of most use.

Have a kitchen chair with a back to it for women who like to sit while they iron or do dishes. You can save time by washing dishes from left to right; also peel vegetables this way. (If left-handed, reverse process.) Do dishes once a day in the evening, stacking breakfast and lunch dishes. When you're building that dream house, install one drawer that can be opened in both the kitchen and the dining-room. On one side, have items needed in the kitchen and, on the other, those used in the dining-room, thus saving both time and space.

In industry, considerable time is spent studying air space—space above. Have lots of shelves and have them built in proportion to the housewife's height. There ought to be sliding doors in kitchen cabinets to eliminate getting hit on the head; also, glass doors so you can see at a glance where things are. Store in low places those things used least often, to eliminate as much bending as possible, which is more tiring than any other exertion.

6. *Children:* Set definite periods of the day for playing with children, particularly babies. They come to expect it and then are more inclined to give you a breather at other times. Have definite place for toys and teach children to replace them after using. With very small babies, have toys in different rooms, so that as you move around with the baby they are available.

Youngsters, particularly girls, like to imitate their mothers. Get small irons and ironing boards, dusters, and baking facilities, and let them work along with you. This is not only educational and can save time for you, but it is also diverting.

7. *Making beds:* The average housewife walks about 170 feet making a bed. This can be cut in half if the housewife will make one side up completely before going over to the other. If no one is coming to view the house, why not just make up the lower part of the bed and leave the rest open? It not only saves time, but airs the bed. When the children get older and are around the house, have one of them help you make the beds. It more than cuts time and effort in half.

8. *Mud room:* If possible, have a "mud room" just inside the back door and containing a sink. Here youngsters can remove boots, rubbers, and

dirt. Also, you can save considerable time by calling in children before dinner is ready, so that they can be cleaned up by the time the meal is served. Many mothers wait until the dinner is on the table, and it gets cold while the kids wash.

9. *Preventive maintenance:* Don't wait until the faucet leaks or the refrigerator goes out of order. Make regular check-ups yourself, or have it done. For a few dollars a year you can have all appliances checked regularly and save many dollars in major repairs. Have a bulletin board in the kitchen to make note of what things are wrong.

10. *Modern equipment:* Any necessary convenience that saves time and energy is not a luxury. It has been proved in industry that by giving workers the best in equipment and conveniences, efficiency increases, as well as interest in the job. Anything that removes drudgery from housework makes for a better home. Sometimes people spend far too much on fancy furniture and things that show and not enough on vacuum cleaners, automatic laundries, automatic stoves, dishwashers, and the like.

Some housewives are loath to discard old-fashioned ways and make use of modern equipment and scientific methods of housework. This is largely because they have, as their guiding genius—Mother—who did things thus and so. All tradition to the contrary, Mamma does not always know best. Sometimes she's clinging to ideas and methods that went out with the old-fashioned icebox.

Take the matter of drying dishes. Some women still make a ritual of it. I know one whose children gave her a new dish-washing machine for Christmas. Although the machine dried the dishes perfectly, she had to run a towel over them herself—to make it official.

The reports in our survey revealed that in most instances women regarded housework with distaste and rushed through it haphazardly, wasting time and energy. Yet, if approached scientifically, it can be not only an easier job even than it is now, but an interesting game.

As I said at the outset, I think the housewife has a "racket," and either doesn't know it or won't admit it. Nevertheless, as an ever-devoted husband, I'd be the last man to begrudge it. And I'm even willing to make the "racket" still easier, if it will make her any happier. Aren't you?

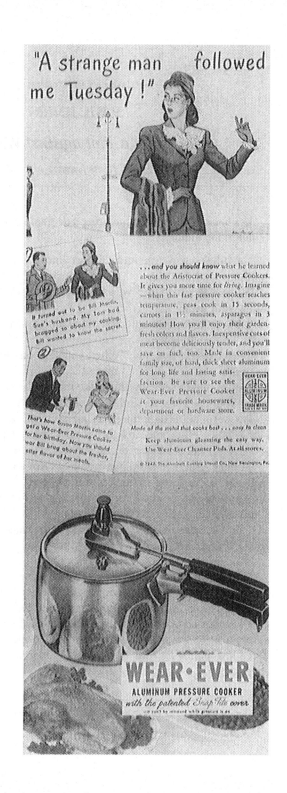

Figure 6. Wear-Ever advertisement from *Ladies' Home Journal,* July 1946.

34

PAUL JONES

"Is There a Plot against Women?"

Ladies' Home Journal, July 1954

Men invent fancy gadgets to bring new freedom to the home. The question is for whom?

The other day we ran into a girl we know, who seemed to have smoke coming out of her ears.

"Why so steamed up?" we asked her. "And what's that you're carrying under your arm?"

"It's a toolbox," she said. "My husband gave it to me for my birthday. Here, let me show you."

"Never mind," we said. "When you've seen one toolbox, you've seen them all."

"You've never seen anything like this one. I won't unwrap it, but I can reveal that the hammer and the screw driver have gay plastic handles. I am not sure, but I think the pliers are set with rhinestones. There is also a little book telling me how to fix leaky faucets, put up kitchen shelves and make a full set of screens and storm windows for the house."

"A very practical gift," we noted approvingly. "We men are often engaged in our larger affairs, and it is a comfort to a husband to know that his wife can do small repairs and general maintenance. A toolbox is something for you to treasure."

"I do not treasure this toolbox at all," she said distinctly. "In fact, I am taking it back. I think it's all part of a big plot."

"A plot?"

"You heard me. You men are shoving off on us women a lot of things you used to have to do yourselves. This morning I was looking at the paper and I saw an advertisement for a wonderful new car-washing gadget. The illustration just showed a hand holding a long nozzle. I need hardly tell you that it was a feminine hand. Somebody is trying to con us wives into believing that washing the family car can be a thrill-packed adventure."

Paul Jones, "Is There a Plot against Women?" *Ladies' Home Journal*, July 1954, 76.

"Surely," we said, "you know the story of the Industrial Revolution. Why should you dream up this plot? For the last hundred and seventy-five years, men have been inventing things, with only one object in mind: to make women happy and carefree in the home."

"All that for poor little me," she said bitterly. "And now I wind up with the toolbox. The more I think about it, the more I realize that men have got rid of a lot more household chores than women have."

"Come, come," we said. "Think of automatic washers and dryers, not to mention vacuum cleaners, freezers and refrigerators."

"That's what I mean. Who used to empty the pan under the old-fashioned icebox?"

"The husband, of course, if he was any kind of a gentleman."

"You see? And who used to take care of the coal furnace? Not the wife. That heavy tramping up and down the cellar stairs, those smothered oaths heard through the floor when the fire went out, were sweet music in a woman's ears. It meant she had a man who was on the ball.

"I am tired," she went on, "of hearing that we have practically nothing to do, compared with our grandmothers. How about you? All you have on your mind now is fiddling with a thermostat or telephoning the TV repairman.

"Actually, grandfather was an even busier character than grand-mother. He put in a sixty-hour week at the plant, and at home, when he wasn't carrying scuttles of coal up to the kitchen range, he was outside, splitting kindling wood. Sunday mornings, after he was all dressed for church, he would go around and wind five or six clocks in the house. You got out of even that dignified duty by inventing electric clocks.

"Besides all that, grandpa had the family rig to look after. He washed the carriage and greased the axles. He cleaned out the stable and oiled the harness and currycombed the horse, not to mention sitting up nights with the beast when he had the heaves. You just put the car in the garage, and that's it. Now you expect us to wash it with that marvelous new gadget I was talking about.

"I doubt if you even know how to cure a ham or a side of bacon. Grandfather did. He had a smokehouse in the back yard that kept him out of a lot of mischief. He was expected to take care of the meat depart-ment, while grandmother put up the fruits, vegetables and preserves. Sauerkraut was also his province. He made it by the barrel, and, in his spare time, chopped up any extra cabbage and fixed a keg of pepper hash. In warm weather, he had charge of the ice-cream department, cracking the ice, mixing it with rock salt, and turning the crank for an hour or so."

"Just a minute," we said uneasily. "You don't expect us to go back to those primitive days, when everything today is so conveniently packaged?"

"Frankly, no," she said. "But I wish you'd quit pretending that modern men are as busy as ever, while modern women have nothing to do. Either that, or get busy and invent a bedmaking machine, a self-operating table setter and an automatic child bather.

"And kindly quit lecturing us on how bad all this leisure is for idle wives, and how we would be all the better for some of the invigorating exercise grandma got over a washtub. I have a good mind to go on a lecture tour myself, deploring the fact that all the evils of modern life stem from man's new leisure, the seven-hour day and practically nothing to do around the house. And now, excuse me, while I trade in this gay little toolbox for a shovel that George can use. It's time he dug up the garden."

35

SELMA ROBINSON

"103 Women Sound Off!"

McCall's, February 1959

Looking down from the head of the stairs into the great ballroom of the Hotel Shoreham in Washington, D.C., you saw ten tables, like ten islands, inhabited by bright, animated women talking or listening or laughing, waving a hand to be heard. At each of four half-day sessions their opinions were recorded by stenotypists. Clustered around the tables representatives of trade organizations, builders, reporters, and even a man from the Voice of America, listened with deep concentration or jotted notes on thick pads.

Most of the delegates who attended this Second Congress on Better Living had been prize winners or competitors in *McCall's* remodeling contests. The rest were chosen with the help of the Junior Group of the General Federation of Women's Clubs and builders whose model houses had been "certified" by last year's congress.

They came from sprawling ranches and from neat New England suburbs, from crowded cities and from towns so small you'd have trouble finding them even on a big map.

Their median age was thirty-two, their median income $8,500. They were the mothers of 233 children—by now 234, since the newest baby made his premature appearance in Kansas City just a few days after the congress.

We asked: "How do you feel about your home and the things in it? Most important, tell us why you want what you want, how you expect to get it."

Nearly all of them own the homes they occupy and more than half plan to remodel. About a quarter of them plan to build or buy a new home larger than the old. Their dream houses are down-to-earth. They like the colonial house best, with the ranch style next. Few like the so-called "storybook" or "Hansel and Gretel" house.

"A good colonial has good resale value."

"I'm a lazy housekeeper so I like a modern house—you can clean it one-two-three."

"Storybook? Icky!"

"I say I want this and this in a house, but I am not too disappointed if I cannot have it."

They like natural, maintenance-free exteriors—redwood, fieldstone, weathered-looking brick. They love white-painted frame houses . . . friendly front doors, good insulation, low-pitched roofs.

"Our stone house is one hundred years old. It's cool when the mercury reaches a hundred. Winter it doesn't need too much heat."

"Shutters are nice. Like earrings. They finish a house."

"Well-designed houses need no ornaments."

"A house should look like a home, not a business building."

They want wiring adequate to the demands of modern appliances. What is "adequate"?

"It means you can turn on everything in the house and nothing blinks or blows a fuse."

More than half of them have installed additional wiring and they'd want a new house wired beyond their immediate needs.

"Wire today's houses like houses of tomorrow."

They'd like builders to furnish a wiring and plumbing diagram to minimize repair expenses.

"And I'd like removable panels in the kitchen and bathrooms so when there's a leak you don't have to break the plaster."

As for windows, they prefer those that open all the way, dislike picture windows staring into their neighbor's picture windows. They are annoyed by cheap hardware.

"If the hardware is poor, so is the whole house."

Virtually all of them want a more spacious house, about eight rooms — at least two and a half bathrooms, including a mud-room half-bath at the rear for quick cleanups; a separate bedroom for each child; a family room (though they do not agree on where it ought to be); and a number of women said wistfully, "A parents' room where grownups can go for quiet and relaxation." They crave more storage space. Said one woman:

"Too many closets would be just right for me."

If they *had* to choose between a small, completely equipped house and a larger one without equipment, they'd take the one that is "all house"; the appliances they'd buy themselves.

"I don't want to keep paying off a twenty-year mortgage on stuff that may be old in five years."

"It's cheaper to add appliances than rooms."

We asked: "What do you think of the model homes offered by builders in your community?" Though they find much to admire, they find much to criticize in some developments.

"Those broom closets they call bedrooms!"

"Too much sameness."

"Georgian outside and stark modern inside."

They like split-levels, but only when they conform to the lay of the lot. They would like builders of development houses to eliminate "gimmicks" like indoor stone or brick planters and sunken living rooms . . . to think more specifically about the family that will occupy the house . . . to leave the trees . . . get some variety into the exteriors besides resorting to trick surface treatments . . . pay more attention to traffic patterns . . . use appliances with recognized brand names.

We said: "Tell us about your furnishings, the silver, china and food on your tables."

Many of them started married life with hand-me-downs from parents and relations. Some bought secondhand furniture; some bought new. But they would tell other newlyweds not to invest in costly furnishings until their tastes mature.

"It's pure providence that most of us can't afford a complete household right away. Now I know what we need and what will wear."

"We started with outdoor furniture in our living room, metal and glass, that was simple and nicely designed. It's in our garden now."

Most of them shop with their husbands for major purchases; some shop around by themselves first.

"I study the outside and he studies the construction."

But they would advise the bride: "Get your sterling and china right away! You may not be able to buy them later."

They are proud of their silver; use it often, some every day; and enjoy polishing it. But most use stainless steel for family, sterling for company along with their "good" china. No matter what their financial status, they consider the evening meal of utmost importance not only because it brings the family together but also because it makes the children aware of good table manners.

"No milk containers or mayonnaise jars on our table."

They like the convenience of plastics but loathe the kind that pretend to be something else. They wish television manufacturers would conceal the screen when it is not in use "so you don't have that great dead eye staring at you all the time." They are familiar with furniture brand names and rattled them off with machine-gun speed. But they want to know more about carpets and curtain fabrics — practical information stores don't give.

"The salesman just tells me how much a carpet costs but not a word about what it's made of or what wear to expect."

"Put that information on tags we can read for ourselves, like McCall's Use-tested Tag."

"Give us a cleaning kit and directions along with the carpet."

They adore vinyl floors, would like cork, distrust asphalt tile. They prefer cotton bed sheets, dislike the feel of nylon. They said no to dark-colored sheeting they were shown, though one woman said, "The brown would be good for children with muddy feet."

They enjoy cooking and occasionally trying new dishes.

"If I had ten maids, I'd still do the cooking."

Their families are less inclined to experiment.

"I had to call kidney stew Cowboy Stew before they'd eat it."

"A surprise is only a surprise if the family likes it."

One woman from a community where families often dine at fine restaurants has "eat-out" nights at home.

"With six in our family, we can't afford restaurants. So we dress in our best, set the table with candles and flowers and sometimes invite friends. The children select at least one new dish from *McCall's* that month the same as they would try something new in a restaurant. They learn about new dishes and it is amazing to see how it improves their manners."

We asked: "What is your opinion of appliances, the ones you have and the ones you want . . . what do you like or don't you like about them?"

On the whole, they are in love with their appliances, particularly the dishwasher and the dryer. They object to the use of plastics in refrigerators, to name plates that collect dirt and grease ("why should I have to clean their advertising?"), to the job of oven cleaning. Many feel that control knobs on ranges should be out of children's reach. They'd like laundry appliances in a utility room off the kitchen. They want their kitchens functional but not clinical.

"I don't want it to look like a hospital or like you stopped for a hamburger in a diner."

"I'm a cook, not an engineer."

Biggest complaint is poor service from dealers. Manufacturers should recognize their responsibility to see that servicing is available. Many said that letters of complaint were not acknowledged.

"An appliance is only as good as its service man."

"The service man read the manual out loud while my husband fixed the dryer. There was no charge."

"At least give us repair manuals written so we can understand them."

Several women paid for their appliances on time[1] because "you get better service if you don't pay in full."

Though they generally disapprove of installment buying or any other kind of borrowing, they feel it is justified for major purchases like appliances, furniture, a car, where immediate and continuing use of the article is worth the interest paid.

These women showed an amazing awareness of family finances. Many of them keep the books and pay the household bills. Some act as part-time bookkeepers for their husband's business. They are familiar with fuel costs, taxes, repair bills. Over 80 per cent of the delegates have mortgages on their homes and they know the size and length of the mortgage, the rate of interest and the balance unpaid.

We said: "Please describe the way of life in your home and your community . . . how you feel about schools and TV . . . about early marriage and children's allowances . . . the use of leisure time."

Their lives are lives of shared responsibilities and fun. Some fish and hunt with their husbands. Some are in business with them or pitch in when help is needed. One quarter have outside jobs.

[1] On the installment plan, that is, in regular (usually monthly) payments.

"When my husband's lumber business was failing, I drove the trucks myself."

"For a farmer's wife like me, there is nothing more wonderful than following your husband on another tractor and feeling you are doing something together."

In turn, husbands help around the house with the dishes, the repairs, with the children, with guests.

Though most of their children are under fourteen years old, 70 per cent of these mothers have already begun to plan for college, many saving systematically for that distant day. They feel that money spent for a college education should be tax-exempt.

"I hope they'll want college, but even if they don't, they are going anyway."

"Our children are small, but we talk to them about college so they'll take it for granted they're going someday. There will be no marriage at sixteen."

Though they agree the "right" age for getting married is a matter of emotional maturity, not chronological age, they take a dim view of marriage under eighteen.

"Some of these girls can't even make up their minds what dress to wear tomorrow. How can they meet marriage?"

"Society prolongs adolescence. Young men can't hope to live an adult life till they are twenty-five if they want to be educated, though physically they may be ready for marriage."

For the most part they are sympathetic to teen-agers, most of whom they find responsible and industrious.

"I think teen-agers are fresh and vivacious. I wish that I was back with them."

"We ought to drop the word teen-agers from our vocabularies."

"Some are spoiled rotten and need a good spanking. But some others are more mature than the parents of other teen-agers."

They disapprove of "going steady," want more planned activities for young people.

TV presents a problem in almost every home that has a set. The mothers of young children screen programs and control the time for watching. They consider shows of violence a disturbing and possibly dangerous influence. "Unhealthy," "distorted," "frenzied," were some of the words they used. Many of the women complained that advertisers have been taking advantage of them by appealing to the children.

"They make my child nag for sponsors' products."

"My children are more affected by the commercial than the violence."

"My children love cowboy shows. I feel sorry for my husband when he gets home with all those horses galloping in the den."

Parents, they feel, should make their opinions known to television stations and to school boards. By and large, these women are not too happy with the education their children are getting; they believe in more homework, less emphasis on grades, more on learning and thinking. They would pay higher local taxes in order to have better schools.

As parents they believe in discipline rather than permissiveness, in a firm hand applied to the bottom when a child deserves a spanking, in teaching him a proper respect for money, property and the rights of others; parents should be parents, not pals. All approve of allowances as the child's right to his growing independence, and chores as a responsibility to the home, not as a job to be paid for.

What will they do when they have more leisure?

"My husband has always encouraged me to improve my education."

"Catch up on my reading."

"I'm studying genetics and I hope as my leisure increases it will lead to a full life when I am elderly, so I won't be a lonely woman dealing cards at Las Vegas."

Still, though they look forward to new horizons, new homes, new ways of living, most of them look back with longing to the parlor, the front halls, the homely comfort of Grandmother's day.

"I love that old-fashioned idea of the family, with a big round table with a lamp in the middle and everybody sitting around, doing homework or conversing. . . . It is that kind of atmosphere I think we have lacked for the last few decades."

36

JOYCE LUBOLD

"My Love Affair with the Washing-Machine Man"

McCall's, June 1960

The All-American housewife is, let us agree at the very beginning, faithful—both to her husband and to her wedding vows. Although she may enjoy reading those novels of suburbia that turn every residential street into a beehive of amorous intrigue, she knows very clearly the difference between that kind of fiction and her own kind of reality. In real life, her job as housewife consists of exactly what the word implies: being wife to a house. And what energy she can snatch from her frenzied round of washing, drying, cleaning, cooking, and circular running she offers directly to the man she loves. And the man she loves is the man she married.

Quite aside from love, we housewives are too busy, too harried, to prance lightfootedly along any primrose path. Infidelity is simply not for us. We don't have the time.

However, at this very moment, I'm carrying on what certainly amounts to a full-blown romance with three men. With each, I am going through all the sweet, stylized motions of a great love affair: the first timid overtures, the long-awaited meetings, the baffling, turning tides of love and hatred intermingled. I'm doing these things; so is my neighbor, and so, very probably, are you.

My men change, from time to time. Some weeks, I fling myself at only one or two; other weeks, I find with dismay that I am fluttering with mad, false gaiety among four or five! Just now, the objects of my affection are three: Charlie Gant, who is tall and thin and repairs my washing machine; Huck Peters, who is short and heavy-set and tends to our television set; and Gus Marston, whose shape has always been hidden by loose-fitting coveralls hung with an electrician's tools. I run winsomely among these three, begging prettily for their favors, thanking them gratefully for their smallest gallantries, dancing to each one's demanding tunes.

I am, let me repeat, a faithful, devoted, and loving housewife. Never-

---

Joyce Lubold, "My Love Affair with the Washing-Machine Man," *McCall's,* June 1960, 42.

theless, I must transform myself into the most flirtatious of plantation belles every time some major appliance breaks down.

Now, you take Charlie Gant. Charlie, bless his sullen soul, is the only man in town who is qualified to diagnose the strange tremors and fevers to which my washing machine is prone. Drawn to his charm or no, I *must* get along with Charlie, since I cannot get along without him. Charlie is a proud man, and he walks the streets of our town with the serious, heavy tread of any despot. For it is he alone who decides between the thrilling promise: "Be over this morning," and the chilling: "Maybe sometime next week."

Charlie very much enjoys the respect and adulation to which he has become accustomed. You can't do too much for Charlie. I mean, if he sees some tendency to grovel, it does not displease him. He likes his coffee black, with two spoons of sugar, and is just the least bit annoyed if I offer him cream.

And so begins the affair.

At nine-thirty-seven of a Tuesday morning, my washing machine gives a single great "Yawp," swirls its load of wash around at breakneck speed for a moment, and then, grinning hugely, disgorges an enormous quantity of gray, soapy water onto the kitchen floor. I dare not waste a minute.

Charlie's phone number is etched on my heart. With shaking hands, I dial, desperate for the sound of his reassuring voice.

Yet it is his wife who answers! My spirits sink. Her flat, unfriendly voice mocks my dismay. Suspiciously, she asks my name. Guardedly, I offer my phone number instead. Then I hang up, forcing myself to be calm, steeling myself to wait.

Now the water oozes sluggishly into the cellar, and mechanically I go about my chores, to the beat of an anxious little song. "When-will-he-call? When-will-he-call? Will-he—horrors—call-at-all?" The water drips on.

The phone rings. It is Charlie! I am overwhelmed with joy. But careful. I must be gay and bright. Charlie hates to hear a woman whine.

"Mr. Gant! Thank you so much for calling back." (Be light! Be bright! Never let him guess how much you need him.) "I ran into an odd bit of trouble with my washer this morning. Really something new this time, Mr. Gant." (Try to be amusing for him. Make your little story entertaining.) "You know, I could almost go swimming in my cellar. It's a leak of some kind, and the water is *pouring right down*—" (Careful. You *mustn't* panic. Charlie wants his women brave.) "And so I wondered if you might like to drop by this morning or—" (Don't push him.)—"this afternoon."

And then his familiar voice, so gruff, so proud, so positive. "Got a lot of calls to make today, Mzz." (Charlie calls me "Mzz.")

"Oh, of course. I know how busy you always are." (Go softly with your flattery, girl. This man knows all the tricks.) "But I did hope you could squeeze me in." (Little giggle.) "I just don't know what to do. I'm just so upset." (*Now.* Cling to him. Forget your pride. This is the moment.) "I really need you, Mr. Gant. I truly do. Any time you can make it will be just fine."

"Yeah. Well, I'll try to drop by. Sometime after lunch." And in his dear abrupt way, Charlie starts to hang up.

"Oh, that will be lovely. I'll be waiting for you." (Little trill of laughter.) " 'By," he says then, with finality. (My, but he's gruff today. It's because he feels he's given in to me. I know his moods so well.)

"Well, good-by, Mr. Gant, and thank you," I say gaily into a dead line.

Well, all this pretty well shoots the morning for me. This kind of thing may have been all in a day's work for the ladies of the French court or for the geisha girl of old; but I have to sit down with my feet up for ten minutes or so, just to get my breath back. I dare not tarry for long, though, because there's a lot to be done to make ready for Charlie. First, the dishes. (Charlie Gant likes to spread his tools on the kitchen table.) Then I make a pot of coffee. (He likes it strong and bitter.) I put the baby down for a nap and the four-year-old in front of the television set. (Kids make Charlie nervous.) It is right about then (hours fly when you are happy) that I hear his ring at the door.

My feet have wings. "Hel-*lo*, Mr. Gant," I carol as I open the door.

He walks silently into the house. I walk behind him. We reach the kitchen and stand face to face. Then bittersweet flows about us. As he stands there, so long-awaited, I find myself in the second stage of the affair. For love is akin to hate, and joy lies closest to sorrow. We are caught in the inexplicable surge of antipathy that lovers know when they try, too hurriedly, to reknit old ties. We glare at each other, helpless in the ever-changing current of our relationship.

"Machine's still full o' water," Charlie says with strong distaste.

"I know it is," I say shortly. "That's why I called you." (The man is a boor. I hate him. Hate him. Hate him.)

"Can't do a thing with the water in her," he says, his toolbox still tightly in his hand.

"Neither could I," I say, fighting this unreasoning hate, unable to keep from going on and making it worse. "Neither could I do a thing with the water in her." There is a terrible silence, a mute struggle between us.

And then *he* gives in. "Well. We can't do a thing till—we get rid of it, can we?" he says as he finally puts down his toolbox.

Ah, the way this man plays on my nerves, the shrewdness of that "we." I melt again; the icy hatred vanishes. "I didn't like to fool with it until you were here," I say archly. "But now you're here to watch the pump, I can turn the timer for you whenever you're ready." He smiles at me, and I smile back.

"I think I can figure her out, Mzz," he says. He smiles at me again and then disappears behind the washing machine. (I love this man. Love him. Love him.)

I turn the timer when he tells me to, and finally the machine lurches reluctantly into its drain cycle as Charlie and I, bound by our extraordinary camaraderie, watch the water slowly recede.

"Appears you've run right through that water-level switch again," Charlie says, and the touch of sternness in his voice chills me.

"Do you think you might be able to do something special with it?" I ask falteringly.

"Dunno. Doesn't look good." (Charlie never really gives in to me. That's the thing about him.) "I'll give her a try, but I dunno," he says, dismissing me.

"I'll get some coffee for you," I say, acknowledging dismissal.

Now he works at his task and I at mine, as a warm, friendly atmosphere comes to fill the room. In time, the job is done. He tells me, in a steady, firm voice, how much I owe. Quietly I get the bills and hand them to him, folded small. (It always seems a terrible thing, somehow, to let money come into our relationship.) He unfolds the bills and counts them carefully, for he never lets my foolish vagaries interfere with what is to him the real cornerstone of our relationship.

"I'll try not to phone again too soon," I say gaily, yet wistfully. (The silly machine may turn on me again before the week is out. I can't let him leave without some whisper of a promise for our next meeting.)

"Yeah," he says equivocally.

The door shuts on his dear retreating form. I sink exhausted on a chair, trying to erase from my memory the entire incident. There is a cry of dismay from my four-year-old. I reach the living room just in time to see our television picture glare briefly bright and then go black. It's all to do again. And with a different man!

5

Fashion and Beauty

Long before the middle of the twentieth century, magazines for women placed great emphasis on women's physical appearance. Nineteenth-century periodicals such as *Godey's Lady's Book* regularly published illustrations of the latest women's fashions, and some magazines that began in the late nineteenth century, such as *McCall's*, originated as a means of promoting dressmaking patterns for home seamstresses. *Vogue* and *Harper's Bazaar* have long been devoted to the world of high fashion. Whereas these magazines have primarily addressed the sophisticated, well-to-do reader, by the 1940s other segments of the population were being given advice on not only how to dress but also how to apply cosmetics, keep skin soft, style hair, and smell good. The magazines for homemakers tended to offer practical suggestions for budget-conscious women who shopped at J. C. Penney instead of Bergdorf Goodman; magazines for younger women, such as *Mademoiselle* and *Seventeen*, focused on the proper appearance for dating, parties, and attracting members of the opposite sex. Whatever the magazine's audience, the underlying message was similar to those regarding housekeeping, marriage, and parenting: Looking good was a duty, requiring hard work and commitment. Titles such as "Do You Make These Beauty Blunders?" suggested just how close women could be to making mistakes and did little to alleviate the anxieties about personal appearance that were also being fostered by films and, later, television.

Even more than in other areas of the magazines' content, articles about fashion and beauty addressed the woman as a consumer. With the exception of occasional articles on exercise (not as highly valued then as it is now), most improvements in personal appearance required *products:* stylish clothing, face creams, shampoos, jewelry, and cosmetics. Articles on the care of hands or hair were published in close physical proximity with advertisements for lotions, polishes, and shampoos. Indeed, the editorial and advertising content of the magazines were often indistinguishable from one another. Long before the term was coined to describe

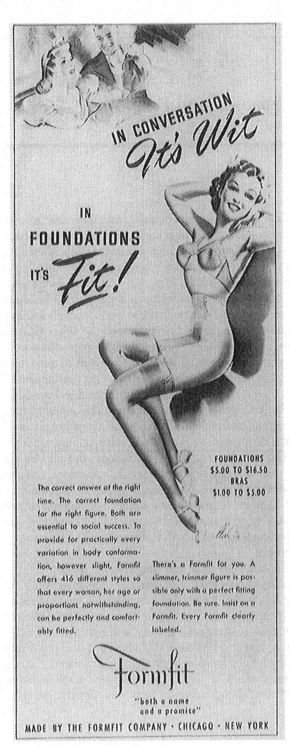

Figure 7. Formfit advertisement from *Woman's Home Companion*, April 1942, 58.

a certain kind of television advertising, the women's magazines featured "infomercials": What was listed in the table of contents as an article about some aspect of beauty turned out to be an endorsement of a particular product or service. The article "Making Less of Yourself" in the September 1955 *Harper's Bazaar*, for example, begins by describing the slender silhouette in fashion that year but is actually an advertisement for a Helena Rubinstein diet and exercise regimen.

The mania for thinness that has helped to create an epidemic of eating disorders was not fully developed in the 1940s and 1950s. In July 1949, *Harper's Bazaar* reported matter-of-factly that the average American woman was 5'3" tall and weighed 133½ pounds and expressed irritation at the fact that stylish clothing was manufactured for women three inches taller. Clothing models in the more upscale magazines were pencil-slim, but the dominant female image in the magazines could best be described as wholesome. There were indications, however, that the idealized body type was undergoing a change during the period. Even a casual survey of the magazines suggests that there were more advertisements for corsets and girdles than for any other single category of apparel. In one sense, this was nothing new: With the exception of a few brief periods, the female body had been molded by such undergarments for many years, usually to make the body fit the fashion of the moment. The pervasive sense of the undergarment ads of midcentury, however, is that the female form needed to be *contained*, even as the ads themselves promised "freedom" from the laces and stays of earlier years. Even more telling of emerging trends was the increased emphasis on diet and weight-loss articles during the 1950s. Whereas a 1949 *Coronet* article titled "Diet Your Way to Beauty and Health" was concerned primarily with proper nutrition, most diet articles of the following decade were intended for weight reduction.

In the area of fashion and beauty, as in so many other aspects of life addressed in the pages of the women's magazines, it is difficult to determine the extent to which the editors attempted to influence women's thinking and behavior and the extent to which they merely reflected existing cultural realities. In terms of the *ability* to influence styles in clothing, hair, makeup, and body type, however, the use of visual images in addition to text could have given these articles a power that is now shared by the television screen.

37

RUTH ANNA READ

"Those Simple Little Exercises"

Good Housekeeping, July 1940

It is difficult to realize that I once led a free, untrammeled life. Incredible that I once leaped blithely out of bed in the morning, bathed and dressed, ate and worked and played, and unceremoniously retired to bed again. Without a thought for the dentist, the doctor, the hairdresser, the manicurist, and the chiropodist. Without a thought for them and their "simple little exercises."

Well, things are different now. They have been ever since the day my dentist broke the bad news. He was in the habit of giving me nasty shocks anyway, and one jolt more or less meant nothing to him. Quite casually, he pointed out that I was on the road to dental ruin and in need of a few simple exercises.

"You see that?" he asked, grasping a mirror in one hand and my jaw in the other. "Oh, it may not amount to a thing," he added, in answer to my wildly questioning eyes, "but I want you to do some exercises twice a day, when you brush your teeth. Take your thumb and forefinger . . . this way . . . . "

Gently he guided my thumb and finger, to show me how to bring back the vigor of youth to my gums. It was going to be easy, he assured me. Just do those exercises every day, and in a few months—or maybe years—my gums would be good as new.

I was only too glad to comply with his request. The exercises took only a few minutes' time, and I was well aware I might be saving myself from the horrors of pyorrhea and what are elegantly termed "artificial dentures." Night and morning I kept at it. I certainly wasn't going to be one of those women who let themselves go to pieces.

The thumb-and-finger technique no sooner had become a routine part of my daily life than my doctor decided it would be an admirable idea to do some simple exercises for *him*. Simple, deep-breathing exercises, designed to buck up my bronchial tubes.

"*In*-hale! *Ex*-hale!" he exhorted me, as together we practiced breath-

---

Ruth Anna Reed, "Those Simple Little Exercises," *Good Housekeeping*, July 1940, 86.

ing in his office. "Now I want you to do this simple exercise every day for five or ten minutes."

As my doctor can tell whether or not I have followed instructions simply by looking me sternly in the eye, I began breathing deeply immediately after leaving his office. Later on, I had to give up this public practice—passers-by mistook me for a victim of asthma; but at home in the country things went very well. Here I could really put my heart, not to mention my bronchial tubes, into the exercise.

In this somewhat stertorous fashion, life continued for some time. Then the hairdresser took up the hue and cry.

"Your scalp's very dry," she told me, in the shocked tone of one who has found moths in the living-room rug. "I think you'd better use our special ointment several times a week. You just rub a little on your scalp and massage it well, like this . . . . "

"Like this," would take only ten minutes at the most, and then if I also did those fifty strokes with the hairbrush which were needed night and morning, eventually my hair would gleam and glisten like tinsel on a Christmas tree. Well, why not? I bought the ointment and felt virtuous, not to say hopeful of the future. After all, it would do me good to get up a little earlier each morning.

With the manicurist who wanted me to rub olive oil on my cuticle each night before going to bed, I also did my best to cooperate. This oiling job took only a few minutes, as the helpful young woman had promised, and if most of the oil came off on my nice, clean sheets, that wasn't her fault. My hands actually looked better, and the sheets were the laundry's worry. I really felt I was getting somewhere, what with the olive oil, the fifty strokes with the hairbrush, the scalp massage several times a week, the daily deep breathing, and the thumb-and-finger gymnastics. Who knows to what heights of health and pulchritude I might have risen, if it hadn't been for that spell of warm weather?

It was spring, and in the spring a woman's footsteps frequently lead her, alas, to the chiropodist. Mine did.

There is no gainsaying that the man worked wonders. I came away walking airily, like a girl of twenty. But I also came away muttering to myself. For the chiropodist had told me my metatarsal arches needed — you've guessed it!—some simple little exercises.

How could I, I pondered, work in these latest exercises? By getting up at six-thirty instead of seven? That would allow ten minutes for the hairbrushing, another ten for the metatarsals, five for the gums, and on Mondays, Wednesdays, and Fridays, I'd need another ten minutes for the scalp massage. Let's see, now. Maybe it would be better to put dinner

back an hour and do my exercises in the afternoon? Or maybe, maybe—
ah there, Satan—I could just forget the whole blooming business? Let
my bronchial tubes worry along with regular grade-B breathing—they'd
done it before. Let my hair grow dull and lifeless, my cuticle dry. As for
my metatarsal arches, they'd get enough of a workout in my garden, without putting in overtime. To heck with the exercises! On the other hand—
How would I look if I went all to pieces?
I confess I still don't know what to do about it. It's a knotty problem. Or
is it—the horrid thought has lately occurred to me—merely middle age?

38

"The Mood Has Changed"

Harper's Bazaar, September 1944

And it's the hat that's changed the mood. The hat is round, and the mood
is a mood of curves. There is a flow, an undulation of line. But it is the
undulation of the tango—strict, narrow, restrained; it stops before it
begins. Simplicity and the slim silhouette are with us still, only there are
no angles. The G.I. look is gone.

It's a mood of fur stoles pulled softly around shoulders, of hands thrust
into billowing muffs, of swinging capes—jacket-length over a suit, or full-
length in fur, or just a little shoulder cape that's part of your coat. A mood
of flowing princess lines in coats, in suits, and in dresses where, though
the waistline hasn't budged, something new has been added: a dropped
sash below the normal sash, a band of fur binding the hips, a fitted
overblouse, a trick, a device to give that strung-out svelteness.

The little black dress you wear to dinner (short, or long this year to
flatter your lieutenant) has dispensed with sleeves entirely, and its slender skirt is full of diversion—panniers or a peplum or a little hike-up in
back secured by a bow—something for movement and swing.

The heavy shoulder pad killed the hat. Now the heavy hat is killing
the shoulder pad. Even in suits, the broad-shouldered masculinity, the
sharp revers are gone. Shoulders are smooth and rounded; lapels are

small, high, pushed-out, and rounded too. And under the jacket your waist is hugged like an hourglass in a waistcoat.

With the heavier hat, the round hat, on your head, you're not the same woman. You're conscious of something lacking at your throat, at your wrists. You'll fasten a dog collar around your neck, or a choker of black pearls and roundels of rhinestones. You'll pull on longer gloves of plaid or striped or plain wool jersey . . . or bold fur gauntlets to balance your fur hat . . . and, in the evening, real creations—long taupe satin gloves, tufted and ballooning over the elbow.

This whole mood of hat, of restrained elegance, will give you a new perspective on color. The all-out, smash-bang combination of the year is brown and black. But you'll feel drawn, too, to winter blue . . . to gray, all the grays from oxford to platinum . . . to "dunduckety" colors like drab and taupe and cocoa and dun (wonderful in satin) . . . and to plaids—real tartans, and charming mongrels in heathery shades of blue, mauve, gray.

39

"Can You Date These Fashions?"

Harper's Bazaar, October 1944[1]

Figure 8 (*pages 200–01*). This "quiz" was an advertisement for Kotex sanitary napkins.

"Can You Date These Fashions?" *Harper's Bazaar,* October 1944,

Can you date these fashions?

Fill in the date of each picture, then read corresponding paragraph below for correct answer.

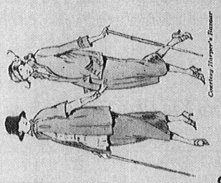

19___

Only bold women bobbed their hair. People cranked cars by hand—sang "Over There". Women in suffrage parade. It was 1918 and army hospitals in France, desperately short of cotton for surgical dressing, welcomed a new American invention, Cellucotton* Absorbent. Nurses started using it for sanitary pads. Thus started the Kotex idea, destined to bring new freedom to women.

Courtesy Vogue

19___

Stockings were black or white. Flappers wore open galoshes. Valentino played "The Sheik". People boasted about their radios . . . crystal sets with earphones. And women were talking about the new idea in personal hygiene —disposable Kotex* sanitary napkins, truly hygienic, comfortable. Women by the millions welcomed this new product, advertised in 1921 at 60¢ per dozen.

Courtesy Harper's Bazaar

19___

Waistlines and hemlines nearly got together. Red nail polish was daring. "The Desert Song". Slave bracelets. The year was 1926 when women by the millions silently paid a clerk as they picked up a "ready wrapped" package of Kotex. The pad was now made narrower; gauze was softened to increase comfort. New rounded ends replaced the original square corners.

Courtesy Vogue

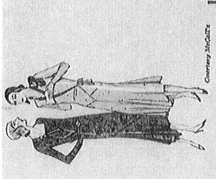

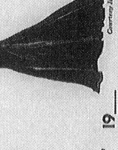

19___ Platinum Blondes and miniature golf were the rage. Skirts dripped uneven hemlines... began to cling more closely. Could sanitary napkins be made invisible under the close-fitting skirts of 1930? Again Kotex pioneered ...perfected flat, pressed ends. Only Kotex, of all leading brands, offers this patented feature—ends that don't show because they are not stubby—do not cause tell-tale lines.

Courtesy MGM

19___ Debutantes danced the Big Apple. "Gone With the Wind" a best seller. An American woman married the ex-King of England. And a Consumers' Testing Board of 600 women was enthusiastic about Kotex improvements in 1937. A double-duty safety center which prevents roping and twisting... increases protection by hours. And fluffy Wondersoft edges for a new high in softness!

Courtesy Harper's Bazaar

19___ Service rules today. Clothes of milk, shoes of glass, yet Cellucotton Absorbent is still preferred by leading hospitals. Still in Kotex, too, choice of more women than all other brands put together. For Kotex is made for service—made to stay soft in use. None of that snowball sort of softness that packs hard under pressure. And no wrong side to cause accidents! Today's best-buy—22¢.

*T.M. Reg. U.S. Pat. Off.

More women choose KOTEX than all other sanitary napkins!

40

SALLY BERRY

"Do You Make These Beauty Blunders?"

Good Housekeeping, April 1944

Sally Berry, "Do You Make These Beauty Blunders?" *Good Housekeeping*, April 1944, 62.

Do you make these
BEAUTY BLUNDERS?

BY SALLY BERRY

Wash your hair but not your brush and comb?

Leave powder to gray your eyebrows?

Wear a skirt with a sagging hemline?

Spoil a neat hairdo with a crooked part?

Tolerate powder smears on collars of dark clothes?

Look spotty with dark powder over light foundation?

Shop for glamour make-up in an old tan raincoat?

41

ELIZABETH POPE

"What Is a 'Well-Dressed' Woman?"

Redbook, July 1945

It would be difficult to find a woman — or a man — who does not understand the implications of the expression "a well-dressed woman," but most of us would be stumped if we were asked to detail the requisites of good grooming.

It is one thing to recognize good-looking clothes; it is quite another thing to describe the intangibles and subtleties of style which add up to feminine chic and smartness.

Professional stylists have definite ideas about the well-dressed woman — they should, because fashion is their business. Here are the opinions of some of the leading designers and connoisseurs of women's fashions.

Hattie Carnegie is one of the most prominent women of the fashion world. She runs a retail establishment in New York for dressing notables of society, stage and screen, and also has several wholesale concerns whose dresses, suits, hats and accessories are sold throughout the nation.

What makes a well-dressed woman, Miss Carnegie says, is an unerring sense of good taste, coupled with a great feeling for utter simplicity and for the best of quality. "She may wear the plainest black dress, but the cut of that dress and the way she accessorizes it, mean everything. Such a woman studies herself in her mirror coldly and then makes the most of what nature has given her, by expressing her real self in the clothes she wears.

"Every woman must first study her own figure. A short woman mustn't wear too much drapery or any lines which cut her height. The tall and big woman must not try to minimize her height, but must give it grace and a fluid line by bias cut and gentle drapery. A big woman must not try to dress like a small one, and a small woman must not try to dress like a tall one."

Elizabeth Pope, "What Is a 'Well-Dressed' Woman?" *Redbook*, July 1945, 32.

Fashions Change — But Be Wise

"The well-dressed woman doesn't fall for fads of the moment in clothes, hair-dos, hats or shoes. She is well aware, however, of the inconstancy of fashion, for she knows that fashion to survive must change. Naturally, every woman who pays attention to clothes is aware of these changes and instinctively adopts them if they are becoming to her. Every great designer, no matter how many tricks she may introduce in a current collection, always adheres to the simple, basic and functional lines of what is becoming to a woman, and the well-dressed woman must follow the same rule with regard to herself.

"Chic must be casual or it isn't chic," says Miss Carnegie. And there you have one of those subtle statements which only women understand.

John Robert Powers, whose Powers girls smile alluringly or are charmingly haughty in nearly every fashion magazine you pick up, of course agrees with many of these ideas. Here are some of his rules for the well-dressed woman:

Dress to suit your type at all times.

Stress naturalness (Powers' School make-up department is now called his make-down department).

Your clothes are important, but you, inside the clothes, are more important.

Correct posture is part of being well-dressed: Keep your head out of your shoulders, your shoulders out of your waist and your waist out of your hips.

Good grooming is essential—never be caught "off guard" with crooked stocking seams or hair out of place. Stress neatness and simplicity.

All Powers girls carry hatboxes for their make-up, shoes, stockings, combs, brushes and whatever else they may need on assignment to a photographer. And inside these hatboxes is a John Robert Powers quotation: "To present a complete and smart fashion picture, a woman should wear a hat."

Hat or No Hat—It's a Question

John Frederics, famous creator of hats for beautiful women and hats to make women beautiful, says, "The best-dressed girl I've seen in a long time was one who had on a mink coat and wore her blonde hair flowing—no hat. Of course if you ask me next week, my choice for the best dressed women might be very much be-hatted."

"Being well-dressed is an instinct," John says. "A woman has to know instinctively what to show and what not to show."

According to Kiviette, one of our leading American designers, grooming is the first requisite of being well-dressed. She first brought beautiful and wearable costumes to the stage in the many musical shows she dressed, and then filtered the glamour of the theater into her day and evening clothes which are sold in fine shops all over the country. "Next," says Kiviette, "clothes should have nonchalance and casualness; they should never look like 'creations.' Once a dress is put on with all the care and attention to details it deserves, it must be forgotten."

Kiviette, a serious artist as well as a dress designer, has some very definite ideas on the use of color in clothes. Here are a few of her general rules:

Blondes should choose cool colors, blues, greens and violet tones. On occasion, browns are marvelous. Pastels as we know them are apt to make blondes fade. White has the same effect unless accented with color. Black is always good, and black velvet is wonderful.

Medium brunettes should use warm color tones, including rich greens and reds. Medium tones and off shades are more becoming than clear, vibrant colors. Navy is often better than black.

Dark brunettes can wear intense colors, as well as yellows and tones of beige and gold. Stark white is wonderful.

Titian or red-haired women look best in greens and blues, and certain terra-cotta shades which harmonize with the color of their hair. Brown and black are also good, but red should as a rule be avoided. Purples, too, unless they are very bluish are apt to clash with red hair.

Philip Mangone, the suit designer, claims you can't write a recipe for smartness, because there are too many intangibles mixed up in it. "The first ingredient," Mangone says, "is that clothes be right for their wearer. "Once she knows her type," he continues, "the one unfailing rule is that her clothes must be fitted properly. An expensive garment that fits badly will not look as well as an inexpensive one that fits superbly. The most important thing for a woman is to have her clothes fitted.

"Next in importance to fit is fabric. A well-dressed woman always buys the best she can afford."

A point of controversy is introduced by Harry Conover, the man whose fashion models and cover girls were glorified recently in the Rita Hayworth movie, *Cover Girl*. He says, "There are two sides to the question of style: a man's idea and a woman's idea. Women appraise the clothes first, then the wearer; whereas men appraise the wearer first and the clothes second."

Conover says that a woman appears well-dressed to him only when she is unconscious of her clothes.

"A well-dressed woman has poise and assurance. She does not fumble or fidget. She does not say in effect, 'Look at me'—you notice her because she carries herself well. Her clothes do not carry her; she carries her clothes."

See Yourself as a Design

Louise Dahl-Wolfe, famous fashion photographer, whose pictures of well-dressed women appear frequently in the fashion magazines, has very definite views on this whole discussion. A well-dressed woman, she believes, has a sense of proportion, sculpture and form. She sees herself as a design, and she selects her clothes and accessories to balance and beautify that design, emphasizing her good points and playing down her bad.

"The attractively groomed woman," says Louise Dahl-Wolfe, "is always neat, with a clean, starched look—there is even a windblown neatness for the outdoor type. Naturalness is not helter skelter; it has form.

"The well-dressed women I have photographed all have a definite sense of balance and proportion. They are able to look at themselves as a design—not just look at their faces and hats and let it go at that.

"They know their good points and emphasize them. A woman with a plain face but a good figure may often be extremely attractive because she is faultlessly groomed, keeps her hairdress and hats extremely simple and wears beautifully fitted clothes which show off her figure. She can wear striking colors in suits or accent a black suit with a touch of brilliant color, but she should never wear a brightly colored or elaborate hat which would draw attention to her face.

"By the same understanding of one's own good points and bad, a

Figure 10. "What is a 'Well-Dressed' Woman?" *Redbook*, July 1945.

Can you tell why this woman is well-dressed?

Tailored felts love little suits.

Trim silver earrings complement her silver buttons.

Softness—even in gloves.

The bag is just the right size for her, matches her shoes.

Nothing clutters up the trim appearance of her suit—the same suit shown on the opposite page: light gray skirt, dark gray jacket with silver buttons.

Her tailored gray felt hat is the perfect companion for her suit: it's the same gray as the skirt.

Her gloves, bag and shoes—all black—accent the soft grays of her outfit.

Fashions courtesy Lord and Taylor

According to the experts, you are a well-dressed woman if—

1. You know your type and make the most of your possibilities. Your coloring, build and the sort of clothes you are most comfortable in help to determine your type.
2. You are always neat and well-groomed, and your clothes are perfectly fitted.
3. You realize that you must accent your good points and play down your bad ones.
4. You use color effectively to accent your personality and the design of your clothes.
5. You are natural and at ease. You are not self-conscious of clothes or appearance. You don't fidget or pose.
6. You carry yourself well. Your head is out of your shoulders, your shoulders out of your waist and your waist out of your hips.
7. You show self-discipline and good judgment in the clothes and accessories you buy. You never buy a garment, hat or accessory which throws your wardrobe out of balance and isn't becoming, anyhow.
8. Your "inner self" shines through and is given expression in your general appearance.

woman with a beautiful face and not too good a figure wears the most inconspicuous and dark clothes with the loveliest hats she can find.

"Above all," says this talented woman who photographs more well-dressed women in a month than the average man sees in a year, "the woman who is well-dressed has perfect discipline in everything she buys. She limits rather than expands her wardrobe, and never buys anything which isn't exactly right for her."

42

JANET ENGEL

"The Fattest Girl in the Class"

Seventeen, January 1948

She couldn't compete with the other girls—not in looks or in brains, she decided. So why try?

When I was fifteen years old I weighed one hundred and seventy-five pounds—on a kind scale. My features were good, but hardly perceptible in the overgenerous setting of my face. I had what the mothers of other—and slimmer—girls referred to, so kindly, as a "sweet expression." I was good-natured—and it's a good thing I was. Good nature was all I had. You see, I was not only big, I was dumb. I'd have been the first one to tell you so.

I'm not sure when I started accepting my own stupidity as a fact. By the time I was in my teens I would have said I was born stupid. I judged that no one gets stupid; you have to be born that way. But looking back now, I can see that I "got stupid"—probably as a child, when I started school. I couldn't do arithmetic in its simplest or most complicated forms (I can't cope with percentages and such to this day) and I was teased about it and frightened. If I couldn't do the multiplication problems that every other child could, the only explanation that seemed to fit the case was that I hadn't enough brains. No one ever set me straight.

Maybe if it hadn't been for my overweight, this idea wouldn't have become fixed. But from the time I was a little girl I had been taken to doctors and treated, without any results. There didn't seem to be anything that could be done about it. I was fat and that was that. In my grammar school days it was a kind of distinction. I was the fattest girl in every class. Maybe someday I could get a job in the circus, I thought without sensitiveness. Fat was my fate, I decided, and that served as an armor against the hurt I might have felt later on, when boys and social activities became important. In much the same way, I decided that stupidity was my fate, too. Nothing apparently could be done about my size; nothing, I reasoned comfortably, could be done about my lack of intelligence. I devel-

Janet Engel, "The Fattest Girl in the Class," *Seventeen*, January 1948, 52.

211

oped a working philosophy: "I can't fight and I can't run, so I might as well be good-natured."

The result? Boys who knew me all of my life were kind to me, but a *new* boy—he'd just avoid me till he found he could tell me his troubles. The girls were all fond of me. I was no competition, and I made a safe, sympathetic confidante for them, too. If Hank told me he was crazy about Eloise, I'd try to convince Eloise he was her dream date. But the chances were that Eloise would then tell me how she dreamed of Joe, the *only* one. This way, I built up a vast fund of intriguing information, and lived on it. "That's good enough for you," I persuaded myself.

As a matter of fact, I got around a great deal, in a secondhand way. I never had to worry about invitations. If Eloise went to a dance with Joe, then Hank would take me, though I represented nothing more than a pair of shoulders to weep on. More than that, he could ditch me at the last minute without a scene.

There was a boy I liked, named Jimmy, but he never asked me. I refused to let it bother me—much, and I told myself that I had to wait till the cream was off the top of the bottle before I got my invitations.

When I was sixteen there was another boy—"the boy"—and he did ask me to a dance. My joy was so great I could hardly bear it. I worried the way all girls do, that my skin would not stay clear, that I would get a cold. I wouldn't even go to gym for fear of sprains and bruises. But I had a worry that other girls didn't have—to find a pretty, young dress in size thirty-eight.

Mother was so pleased to see me really interested and happy about myself that she spent days shopping without me till she found the dress. We thought it was perfect. It was red chiffon, and the skirt was scattered with ostrich pom-pons with rhinestone centers. When I walked and set the skirt in motion, it seemed like a star-studded cloud.

The night of the dance mother lent me her red velvet evening wrap that matched the dress perfectly. It was lovely. "He" came, and I forgot for that moment that I wasn't lovely. I made an entrance, slowly, down our circular staircase to the black and white marble hall below—a movie entrance. "He" gasped, and as he came over to me he took my hand and whispered in my ear, "Janie, God made the little bird red and the elephant grey." The castles came down; I was completely shattered.

Well, only a stupid girl would have fooled herself that way. I told myself I'd forgotten my limitations. I couldn't wear red, or have a beau, any more than I could solve an equation. So I sat back and relaxed a little more. And I smiled when I didn't know the answers. "If a train traveling east went at the rate of fifty miles an hour and a train traveling west went at a rate of sixty miles an hour, when would they meet?" I hoped to God never. Let someone else worry about it. I'd try only the things I could do. One of the things I could do was run a party. When there wasn't enough money in the class treasury to have a party, the problem landed in my lap. I'd think of the simplest solution because I knew if it were complicated, others would lose interest. I'd have everyone bring two lemons for lemonade, and a box of crackers. I'd borrow a victrola,[1] and there was the party. I'd manage everything and it would be a good party. But all that took was understanding other people. And that I did. I knew that if everyone chipped in, everyone would feel it was his party and really work on it. I knew that once the boys get on one side of the room and the girls on the other, a party's done for. I never let it happen. I never let one girl snatch all the really terrific boys, nor would I let any girl be alone for any length of time. I knew what it was to be a wallflower. And I learned a lot about people, just being one. I did have horse sense.

But no brains. Every day made me more sure of that. Like the day we had a mock democratic convention at school. I represented the State of Utah. When the roll was called, all the other students representing other States had the brains to vote the way they thought the teacher would like them to. Then they got to me, and I announced that Utah's six votes were cast for the other, the unorthodox side of the question! The assembly roared. A terrific wave of whispers swept the room. I got up and made a speech on the subject. And all the other States called on after me voted my way! Afterward I went around thinking what a fool I was. The fact that I'd turned the school vote seemed to me more stupidity, not leadership.

A day or so later I was called to the principal's office. Mr. Schorling was the kind of man whose very look inspired fear. He was strict, unsmiling. None of us really knew him. I shook on the way in. My vote and speech at the mock convention must have been even more serious blunders than I had imagined.

Mr. Schorling told me to sit down. Then he said bluntly, "Jane, I'm afraid we'll have to ask you to resign from school!"

I stopped breathing for a shocked moment; then I stammered that what I'd done couldn't really be that bad and couldn't I have another chance.

Mr. Schorling said he didn't know what I was talking about. "It's not what you've done, it's what you haven't done. Why, the only achievement to your credit since you've been in this school was the very good speech you made the other day. You're lazy, Jane. And we can't afford to have

[1]A record player.

lazy people in the school. Your marks are next to the lowest in the class and every quarter you do a poorer job."

I felt comfortable for the first time since I'd entered the office. "Mr. Schorling," I said, "if that is the reason—my marks—it's simple. I can't help my grades. I'm stupid."

Mr. Schorling looked at me curiously. "I don't know where you got that idea, Jane," he said slowly. "You are not stupid. We have methods for checking the intelligence of our pupils. Your intelligence grading places you in the upper third of the upper third. You can think—our testing proves that. What's more, you can think on your feet. I saw that the other day. But you're not trying. You are lazy."

I did what anyone else would have done in my place. I started to cry.

"I'll tell you what we'll do," Mr. Schorling said. "We'll give you a trial for another semester. Then we'll talk it over again."

When I left the office, I was terribly depressed. What would my mother and father say if I were expelled? And I was confused, too—painfully confused. Mr. Schorling's words kept ringing in my head . . . "the upper third of the upper third . . . ." It couldn't be. Maybe I didn't study hard—but what was the use? I didn't have plans for a career or for college like the other girls—how could I if I wasn't capable of carrying them out? But if what Mr. Schorling said was true . . . was a fact . . . . There must have been some mistake. Maybe I had made a good speech. But that was just my horse sense again. Anyone could have seen what was wrong with the voting that preceded mine. That wasn't like learning the reasons for the Civil War, or figuring out the length of a shadow cast by a six-foot tree when the sun—Oh, I'd try—I had to—but I knew I couldn't do it.

I worked and worked . . . so hard I even lost a little weight. Then, almost before I knew it, came the next semester. Again I was summoned to the office. My knees were so weak I thought I would have to be carried, but I made it. The ogre sat behind his desk, unsmiling.

"Well, Jane," he said, "what do you have to say for yourself?"

"Nothing," I answered. "I tried."

"Great goodness, girl," he said. "Couldn't you feel, can't you tell the difference? This semester you are second from the top of your class. You are not stupid—you never were. You've just been lazy. Stupidity we can forgive, but not laziness. From now on, you work or stop school."

I left his office this time with an exhilaration, a new self-confidence. I had brains and I'd use them.

I suppose Mr. Schorling knew that "laziness" was a simplification of what ailed me. Probably it was apparent to him, then, as it is to me now, that I was really running away from the challenge that my size imposed.

A big fat girl has a tough job cut out for her if she wants to be attractive and popular; she has to forget what she can't fix and make the most of charm, personality and intelligence. I had been afraid to try—probably fearing I wouldn't succeed even if I did. I just gave up, tacking on the added excuse that I was stupid, resigning myself to being "good old Jane" and making no attempt to develop any of my genuine capabilities. I denied having any. I get chills now when I think that I might have gone on that way indefinitely—if Mr. Schorling hadn't brought me up short and given me a target well within my range. Apparent stupidity I could overcome. Studying would do it. I didn't worry beyond that first step.

But one thing led to another, of course. The more I worked, the better I did in my classes. My self-confidence grew and I began to see that I could hold my own. Opinions and ideas aren't momentary accidents, but a part of a personality. Moreover, they spread like mint in a garden—give them a chance and more ideas flourish. I began to be a person, not just a sympathetic reflection of my friends. My stock went up . . . with them and with myself. I began to plan for my future.

Over the years that followed, I did lose my ungainliness, although mine isn't a story of a transformation into a slim beauty. It's just that what I looked like became less important than what I was. Work took off some of my poundage; good taste in what I wore minimized the rest. And it didn't matter. What mattered was knowing my abilities, using them—and sometimes even giving them a too hard job, just to keep them in trim.

43

"The Lass with the Delicate Air"

Mademoiselle, July 1949

It seems painfully rudimentary to be saying that no attractive girl ever has an unattractive smell about her. For so many years ads have described, menacingly, the social suicide associated with the faintest taint of perspiration odor—no friends, no beaux, no success, no nothing nice. Strong stuff, but it brought results. Today practically all women, and many of the less self-satisfied men too, use some corrective preparation in order to be completely presentable to themselves, to one another and the world at large. Good thing, too, we all agree.

Then why bring it up at this late date? Simply because three separate people mentioned, only recently, that the products they used were not really satisfactory. Delving into the causes, they turned out to be: the first girl was using hers all wrong, just never bothered to take the directions seriously; the other two were relying on products much too mild for their particular degree of glandular activity. All of which led us to believe that there's still some light to be cast on the subject of perspiration control, and certainly on some of the fine points of product performance and suitability to the individual. It's not the simplest subject to present either prettily or completely; we'll try not to be too dull.

At the very beginning, then, there's the bath. A bath washes away every trace of stale perspiration, gives a fresh start. But unless you literally live in the tub, soap and water alone have no inhibiting influence on either perspiration or perspiration odor. As soon as you've toweled your skin dry, the machinery starts up again. It's sweat glands, those millions of tiny thermostats giving off their daily salt-water output, usually over a quart, sometimes as much as three. This sweat itself has little or no odor when it is fresh. Other glands, though, are in business, discharging fatty acids, urea, acetone and nitrogen, and when these and the bacteria already on the skin or in the air all combine, decomposition sets in, odor develops again.

It's possible that if you never ate onions, garlic, spicy foods, never smoked a cigarette, never drank a cocktail, and bathed six times a day, you'd never need either an antiperspirant or a deodorant. And even then, you couldn't be sure, because there are other factors, as diverse as emotional shock, an ordinary hot meal, jumpy nerves, endocrine imbalance, overweight, certain illnesses and the very medications prescribed for those illnesses. (The smell of ether breathed in by a patient before an operation is breathed out by her glands for hours afterward. And just as the smallest amount of alcohol is easily detected on the breath, it can often be noticed secondarily in perspiration.) Again, glands are most active during physical effort or exercise, in hot weather, just before and during menstruation and when clothes are too heavy or too tight for the climate and comfort. Perspiration and odor aren't uniform, for even in the same body different areas vary greatly. But the underarms, the most noticeably troublesome areas, are really handicapped; they have especially large sweat glands and are generally covered closely by clothing.

So, seeing the odds against complete security for the average person, the idea is to find the right product, discover the right number of times to use it, and—terribly important—apply it in the right way. We wish that that first requirement could be as easy as it sounds. Actually, a corrective that means pleasant, comfortable immunity from personal untidiness to one person can mean a skin irritation to another. That, or the opposite extreme—it can be completely ineffectual. But there *is* a perfect product for everyone who'll take the trouble to experiment, even for people who have extra allergies to call their own.

Concerning the categories of these correctives and what they can do: There are still single-purpose preparations, deodorants that prevent odor, antiperspirants that check perspiration, period. Today, though, practically everybody wants miracles, which in this instance means for the same product to do both. And while it has taken years of research, it's finally possible to choose from dozens of products, even in several different forms, all offering protection from *both* perspiration and odor. Some give more, of course, some less, depending partly upon their formulation, partly upon the user's own physical make-up. Their trade name is, none too gracefully, deodorant-antiperspirants.

Now, the antiperspirant half of that name means exactly this: an astringent action contracts the openings of the sweat glands, diverting all but a fraction of the perspiration flow (maybe only one per cent) from the underarm area. Where does it go? To other, more exposed parts of the body where it is given off in a normal but much less annoying way.

The active ingredient that provides this astringency is usually one or another of the aluminum compounds, aluminum chloride or aluminum sulphate, sometimes alum or zinc oxide, and we drag in these drab laboratory words because your own experience will indicate exactly this: that you *need* and can use, easily, a product incorporating such an astringent ingredient, or that your skin won't tolerate a product containing it. And, on the same principle that if you're sensitive to strawberries it doesn't take a bowlful to bother you, a skin sensitive to an aluminum compound won't accept it placidly in even the smallest amount. How are you to know, then, of its presence? Look on the container, where Federal regulations and manufacturers' ethics demand it be printed.

In general, the cream types are milder toward a sensitive skin than the liquids. They contain less astringent and, since astringency is what checks perspiration flow, a cream is effective for a shorter period. No problem there, though—just use it oftener, that's all. With most people this means once a day. That's nothing, really, for creams have the advantage of being easy to apply. They seem to vanish soon after being stroked on the skin, with no required wait for drying.

Liquid types are especially good for those of us who perspire more and need stronger discipline. You don't kid around with a liquid deodorant-antiperspirant worthy of its name. You *must* wait for it to dry, you *must* rinse it thoroughly from the skin with clear water and pat dry with a towel. And the longer such a corrective stays on the skin before the rinsing, the longer its effect will last. To give an example: one of our favorites in the dual-purpose liquids offers this ratio of control: letting it stay on the skin three minutes before rinsing off means safety for one day. Fifteen minutes, pre-rinsing, takes care of two days. And putting it on at bedtime, rinsing off under the morning shower, is preventive for anything from three days to five, depending upon weather, nerves, et cetera. The directions allow for shaving immediately after use of the product, but specify a wait of one day *after* shaving before another application. Further, in the overnight method, this manufacturer specifies a pause of several minutes before putting on what he terms night attire, saying it's advisable to protect delicate fabric from contact with the armpits. For people who read themselves to sleep, like ourself, the drying time goes fast enough. A murder mystery propped against the knees, the hands resting, clasped on top of the head, and before you know it . . .

In charting the imperatives connected with the use of a strongly effective deodorant-antiperspirant, we've made it sound like a good bit of trouble. It's not. As one who constantly guinea-pigs new products in the name of the *Mademoiselle* reader, we've found that the small extra amount of care connected with the use of a liquid deodorant-antiperspirant is justified, ten times over, by the added protection it gives at such times as it's needed.

You, of course, are free to use the same product continuously, once you've found the one best for you. Good idea, for then you're more familiar with its performance and can schedule its use more efficiently. Any number of people tell us that their routine is, like clockwork, Monday, Wednesday, Friday. What could be simpler? Hotter weather and other emergencies can be coped with as they come up, but it's smart to maintain a skeleton schedule as a basis of operation.

Some liquids now come in squeeze-spray bottles, and very nice too. They spray a fine mist direct to the underarm, drying time is considerably lessened, otherwise the rules of the game are much the same.

We seem to have concentrated on the liquid dual-control products, but since they need the most explaining, apply to people who need the strongest measures, that's only fair.

Let's take other cases now. When perspiration isn't excessive or when dress shields[1] are a part of wardrobe and habit anyway, the main concern is to be free of odor. A simpler, milder product, then, can give all the protection required. It can be a deodorant with only a minor antiperspirant action, or a straight deodorant itself. These come in every possible form: powder to shake on or dust over the skin, stick or compact to stroke on, milky lotion to apply with fingers or cotton, clear liquid in combination with cologne (some that exactly match your favorite perfume, yet), saturated pads of flannel, pleasant fluffy creams and, before this sees print, there may be new ones.

Some of these products, the milder colognes and powders especially, make excellent hot-weather accessories for the person who uses, underarm, the stronger astringent types. In summertime particularly, it's comfortable and convenient to exert partial control over perspiration on feet, palms and between shoulder blades.

The effectiveness of all these varieties is, again, variable with their formulation and your personal chemistry. But beyond a doubt there's a right one for everybody.

Now about fabrics, have you ever seen what unchecked perspiration alone can do to a dress — the staining, the permanent discoloration, the ultimate deterioration of the fabric itself? Then there should be no shocked indignation on finding that any chemistry formula capable of *preventing* that perspiration can do the same damage — if application is careless. So never, if you use a liquid deodorant-antiperspirant, step into your clothes without first rinsing and drying the underarms. Personally, the only way we deviate from directions is to always be more careful than they specify. Even with cream types, which don't usually demand this, we always go over the treated surface a little later with tissue or a damp washcloth. That removes any chance of excess cream working its way into the fibers of fabric that will come in contact with it.

Another fabric fact we'd rather have you read here than learn sadly: between dry-cleanings or launderings, never press a dress that has been stained by perspiration or by an improperly applied perspiration product. The iron's heat sets up (and seals in) a chemical action producing a sickening stale-sweet odor and possible damage to the material itself.

Tracking down the right perspiration control product is scarcely a fun process, like finding the right lipstick shade or the prettiest hairfix. It's one beautifier that doesn't show, one you can't possibly do without. It's

[1]Dress shields were pads worn inside a dress or blouse under the arm to absorb perspiration.

there, all right, it's up to you to find it—and follow directions. Then you can always be fresh and fine.

44

"At My Age"

Harper's Bazaar, December 1949

Collected from smart women of fifty and up, these tips on dressing your age cleverly:

"You can wear good clothes for years and years, but never last year's hat. I find I look my youngest when I wear as little hat as possible—a turban (new again), a pillbox, any version of the becoming beret with a veil."

"Corsets and coiffures—that's where I put my money. The best dress in the world looks like a rag if your girdle fits badly. And all of us look like witches if every hair isn't in place."

"I liked black when I was twenty and I still like it. But now I find the rich black of velvet, the crisp black of taffeta or the soft black of chiffon much more becoming than flat black crepe. With white hair and high color, sharp bright shades—poppy red or purple—are extraordinarily becoming, and much much smarter than maroon, plum or sage green."

"The arms may go, but a good décolletage lasts almost forever. Just because you are sixty, don't feel that you must have a genteel semi-demi neckline. All my evening dresses are cut dashingly low in front, though the sleeves are long."

"If I leave my suit jackets unbuttoned, and wear an unbuttoned cardigan over my sports dresses, I find I have a better line around my middle when I come upon a mirror unawares."

"Pass up the junk jewelry. Instead, lump the loot of your lifetime and have the stones set in one or two really handsome pieces."

"No fussy jabot blouses in my closet! I prefer fine plain silk shirts—creamy marocain, pale blue satin—and at this moment I'm off to shop for one of the new chiffon pullover sweaters . . . to wear with my gray suit."

"Ask me what I want for Christmas—a *black* mink coat, much more becoming to my white hair and blue eyes than the brown."

"At My Age," *Harper's Bazaar*, December 1949, 118.

45

"The Line Forms Here"

Mademoiselle, July 1952

Some of the best figures on the beach are invisible in street clothes, while many a so-so body, when you put clothes on it, will stop traffic. This may be justice of a kind, but we think there's a more concrete reason behind this state of affairs. Ten to one the made-over beauty, who can't take her charms for granted, is wearing a girdle and bra that understand her.

If you can get by without this small assistance you're a rare and lucky girl, but why just "get by"? The girls with the dazzling skins are the first to check their mirrors and pat on powder. The girls with the lovely long legs are the girls who understand best the cosmetic value of nylons. The girls with the celebrated figures—models, ballet dancers—wear some kind of girdle on the job as a matter of course. For a foundation can be the difference between a good figure and a spectacular one.

There's another reason this fall why a girdle is going to be more than a prerogative or a place to hang your stockings. The word is out: "A woman will have to do something for her clothes again—not count on crinolines and buckram to provide the figure." Last year we had a holiday: all you needed under your full skirts was a waist-nipper and a petticoat. This year there won't be any secrets about your figure. Look at the fashions. The newest ones are slim. They're supple. There are no hard and fast lines and we aren't suggesting any hard and fast corseting. While your figure has to be slim, it shouldn't and *needn't* be wooden. A little deft remodeling here and there is all. Fabrics are softening up and this means that clothes will take on the shape of the body, both where they're fitted and where they merely cling. What fullness there is, is a soft fullness and often starts low on the hips.

There's a flock of new dresses that are vaguely Empire, plastered to the body from just under the bosom to the top of the thigh and reporting every nuance of every curve. Unhappily, if you're wearing a girdle roll or an ice-cream roll around the waist they'll report that fact too. The charm of these dresses is the flow of the figure, the clean drawing of the curves. To get that flow your girdle must travel smoothly up the figure. Your bra needs to be high, smallish and roundly cupped. Or you might try one of the new light and airy all-in-ones—no relation to the all-in-ones your mother may have worn way back when—made without bones and actually pretty.

Then there are the middy dresses. They ask for understated bosom and hips. This doesn't mean you have to tape yourself down but it does mean a bra that minimizes your bosom, a girdle that gives you a firm pat on the back of the hips.

Waistlines will be worn in no uncertain manner, swathed in cummerbunds or embraced by hearty belts that go from high to low. If you have a respectably small waist to start with, a girdle with a high, smooth, non-collapsible band at top will make it look *incredibly* small. If your natural waistline is neither here nor there, the same girdle will shape one for you.

Many suits have narrow box jackets over still narrower skirts. If your bosom is generous it takes a talented bra to keep the box jacket from looking like a boxed-in chest. The worst thing that can happen to a slim skirt is a thigh bulge or a garter bump. A girdle should ride smoothly past the curve of the thigh or be cut away in front. Before you buy, sit in it and see what happens.

More and more, sweaters plain and fancy are incorporated into clothes. The only way to buy a bra to wear with sweaters is to try it on *under* a sweater. Bend over and fall into the bra rather than yank yourself into place. The correct separation is as important as the correct size. Check to see that there's no binding in the back making an unhappy puff of flesh there, and adjust the straps.

There, now. Didn't hurt at all, did it!

46

BERNICE PECK

"Accessory after the Body"

Mademoiselle, October 1952

A body isn't necessarily a figure. A body is what you've been given, a figure is what you make of it. There are plenty of accessories (you could easily call them that) for helping you shape this pretty transformation: the right clothes over the right girdle, the right bra. And other even more fundamental accessories: the right diet and the right few exercises. The combination of all these can turn a body—anybody's body—into a real figure.

Doing this can be a lot of trouble or a little, depending upon how close to the ideal are your own given proportions, how intelligently you approach your problems and how lucky you may be in finding a method that will make you want to finish what you started.

There's a new way, perfectly healthful and quite easy, easier than *any*thing up to today's date, for making alterations on a body. It's a way worked out by that woman remarkable for making many women handsomer and happier—Ann Delafield. It's a safe method because Ann Delafield is as much concerned with health as she is with beauty, and works closely with many doctors. It's an easy method because Ann Delafield knows the weakness of the flesh (especially when there's too much of that flesh) and so devised it to *be* easy.

The basics are a capsule and a wafer. The capsule is a vitamin supplement, scientifically balanced. The wafer is an appetite reducer, something you put in your mouth when your will power deserts you. It tastes like candy. Unlike candy, it's packed with eight or nine valuable food elements, all low in calories, all high in strengthening proteins, minerals, vitamins—not a fattening thing in it. Taken when body energy and resolution sag low, the wafer is a small sweet meal that stills the pang of hunger so disastrous to the best dieting intentions.

You'd like to know what you eat, of course, aside from these two miniature fortifiers. You eat enough and you eat very well. For instance:

Breakfast gives you fruit, toast, eggs, coffee. You lunch on fowl or fish,

Bernice Peck, "Accessory after the Body," *Mademoiselle*, October 1952, 87.

a vegetable, fruit, coffee. Dinner offers meat, potato, two vegetables, coffee and even cake. So what have you got to lose? Nothing but what is extra, at the unsuffering rate of two to three pounds a week. If you choose to lose at an even faster clip (and only if your doctor says you should), there are two alternate diets with, naturally, less to eat.

There's even more to the Delafield plan than meets the palate. Other aspects of personal improvement are there too, these in her brightly illustrated book, the *Ann Delafield Appetite Reducing Plan,* which comes with the thirty-day supply of the capsules and wafers for $6.95. The book gives fluidly easy exercises, important points on skin care, make-up manners, basic hair styling for your particular brand of face—and gives all this information concisely and without unnecessary frills. It's the very essence of simplicity for helping you to help yourself to *all* the accessories of health and good looks. The entire program is so sensationally good, as well as so simple and so inexpensive, that if you are interested in shaping a beautiful figure from whatever you've been given for a body, you'll find this the really enjoyable way to do it.

47

"Making Less of Yourself"

Harper's Bazaar, September 1955

She's as slender as a willow wand, this year's lady—and the less of her there is, the better. For one thing, it's good medicine, being in good shape; she feels fine, and looks it. And so does her new fall wardrobe. Observe the cut of her clothes: the narrow suits, the narrowing coats, the narrow, narrow dresses. That's *her* figure they're echoing—the small, high bosom, the whittled waist, the nicely diminished hips. A gift from heaven? Sometimes it is. More often, it's the result of a regime. And that, according to Mme. Helena Rubinstein, means more than the simple admonition to stop eating. The Helena Rubinstein program for the good loser is, in fact, the first really complete reducing plan to be found outside the rarefied purlieus of the beauty salon. It copes with your menus, yes, and with your state of mind as well. It begins with a "Reduce Book," with some useful tips and four flexible diets worked out on the logical assumption that few women can manage an ironclad food schedule. For the nibblers—those temperamentally inclined to borrow from Peter to pay Paul—there are pleasantly flavored tablets, practically calorie-less. And finally, to preserve you from the frequent aftereffects of a real weight loss—the drawn look, the muscles that have forgotten their job—there is Body Firm, a massage cream (smells delicious, too) and Lithe Line, an easily manipulated exercise rope to pull you together. These are the ingredients; further details about each of them are to be found elsewhere in this issue. Faithfully followed, the whole plan should turn you into a scaled-down Eve . . . .

48

ELINOR GOULDING SMITH[1]

"How to Look Halfway Decent"

McCall's, February 1959

In order to be truly beautiful you must study yourself and decide what kind of woman you are, and then work at bringing out your own personality and character. Are you the sultry type, the tall, languid type, or the vivid, dynamic type? Perhaps yours is the wholesome-American-girl kind of beauty, or the pale, ash-blond, fragile kind. You must decide now, and plan your make-up, your hair and your wardrobe to enhance this style of beauty. I carefully analyzed myself and decided that my real, true, individual personality was the artless, instinctive, clean-cut, wind-blown slob, and the whole thing began to get much easier.

You are well on your way to being utterly devastating if you can develop a serene belief that you *are* utterly devastating. This fools people into

---

[1]Elinor Goulding Smith was the author of a number of books poking fun at the impossibly high standards the culture set for women's performance as wife, mother, and homemaker. Among her books were *The Complete Book of Absolutely Perfect Housekeeping* (1956), *The Complete Book of Absolutely Perfect Baby and Child Care* (1957), *Confessions of Mrs. Smith* (1958), *The Battered Bride: Things Your Mother Never Told You* (1960), and *Elinor Goulding Smith's Great Big Messy Book* (New York: Dial, 1962).

Elinor Goulding Smith, "How to Look Halfway Decent," *McCall's*, February 1959, 56.

thinking you're really beautiful, which is very likely the secret behind Marilyn Monroe, Gina Lollobrigida, Brigitte Bardot and Audrey Hepburn, all of whom only *look* beautiful. This belief in your own beauty is probably the hardest part of the whole thing. As a matter of fact, it's the one that I got stuck on. You have to work at it. The way you do it is this: You get up in the morning, you stand in front of the mirror and you look at yourself and say, "I am *gorgeous.*" (Of course, if you're just going to stand there and giggle, I can't help you.) You must practice this until you can stand calmly and simply not believe what you see.

Being absolutely stunning also requires attention to details of grooming. Women like Marlene Dietrich, Dolores Del Rio and Elizabeth Taylor probably got by just on good grooming. This one isn't easy either. The idea is to look at all times as though you have just stepped from a shower and dressed in clothes designed for you in Paris, even though you have just finished cleaning out the garage in a 1943 matched set of dungarees and arctics. I have this gift, only in me it's in reverse. I step from the shower, do myself from head to toe, and look as though I've just come in from a long, hard day in the potato fields. Oh, well, it isn't everybody who can do that.

Beauty is a day-in, day-out job to which you must devote yourself. Some women think they can run a comb through their hair, spray it with lacquer, throw on a little gold hair powder, slap on some astringent lotion, foundation cream, eyelid cream, neck cream, powder, rouge, lipstick, eyebrow pencil, eye liner, mascara, beauty spots, eye shadow, hand lotion, invisible chin tightener, nail polish and perfume and have done with it; and out they go into the public gaze with their eyebrows unkempt and a thread on their skirt. Naturally this sort of careless, hasty, haphazard effort is completely wasted. Beauty is not for them, with their unbrushed eyebrows.

True beauty starts with sparkling good health, which means that starting right now you must resolve to get eight hours' sleep every night. Give up smoking, drinking, spices, fried foods, high heels, coffee, tea, cocoa and yachting. Put yourself on a diet and lose or gain thirty pounds—never mind what you weigh now. You just want to *feel different.* Don't forget that you're bringing out a new and exciting you. Your daily schedule must leave you time to lie down and rest with eyes closed and lotion-soaked pads on your eyelids.

During this rest you must relax completely and forget cares and worries. Think lovely thoughts. *Let* the baby cry. Let the four-year-old put his *own* bandage on his knee. Let the eight-year-old *walk* to Cub Scouts—it's only six miles. *Let* the ten-year-old skip his piano lesson. Let the fourteen-year-old figure out his *own* algebra problems. *Let* the sixteen-year-old elope with the unemployed tuba player. *You rest.*

It is also helpful to set aside a half hour a day in which to do exercises. You must decide which exercises you need most, but if you're not certain which you ought to do, do them all. They're bound to do *some* good. Do exercises to slim the ankles, the hips, the thighs, the upper arms, the wrists, the waist, the neck and the fingers, the shins, the forearms, the scalp. My trouble is that I need an exercise to shorten my chin and I haven't found a really good one yet.

Rouge, cunningly applied, can give the illusion of widening a narrow face or narrowing a wide face, shortening a nose or chin that is a shade too long, making the cheekbones higher or lower. There is hardly a defect that cannot be concealed or minimized by the careful and subtle application of rouge. I tried this technique once. I put a tiny bit on my chin, which is a trifle too long, and a small amount on the tip of my nose, and the least touch on the bump of my nose where it was broken, and just a shade of rouge on my cheekbones, and the tiniest touch just under my eyebrows to make my eyes seem brighter, and a minute spot at the inner corners, and a speck at the outer corners of my jawbones, where they might be a little bit narrower. (The effect never seems just right, but it is possible that I don't apply it cunningly enough.)

Beautiful hair is a positive must, but beautiful hair doesn't just happen. You have to *struggle* with it. Hair should gleam like bronze or copper or gold, and not, as in my case, lie around like crab grass. It must be shampooed and brushed constantly, and it should have permanent waves, sets, curls, and be sprayed regularly against brittleness, dullness, softness, stiffness, tent caterpillars . . . . Oh, hair needs a lot of care. Of course, as I'm aiming for a different, more individual type of beauty, I don't have to bother. Mine just lies around and does what it pleases.

Hair styling, too, can conceal tiny imperfections in face structure. Hair full at the neck can help a narrow chin or jaw; hair swept back will make a wide face seem less so; soft curls on the cheek can conceal or soften a square jaw line. Only careful experimentation and the advice of a good hairdresser can help you to decide on the most becoming hair style for your own special kind of beauty. I cleverly conceal a high forehead with bangs. Now I'm looking for a way to wear bangs that will hide my nose and mouth. Actually, though, I shouldn't complain. I have only three slight imperfections that keep me from being a true beauty—my hair, my face and my figure.

When after all your efforts you at last become beautiful, glamorous and lovely, you'll enjoy it. You really will. As for me, I'll just go on working at the—well, let's call it *character*—in my face. I'll be different and let it go at that.

6

Critiques of the Women's Magazines, 1946–1960

Criticism of American women's magazines began long before Betty Friedan, in her 1963 *The Feminine Mystique,* chided these periodicals for pressuring women to expect fulfillment in the role of homemaker and establishing impossible ideals for their performance of this role. As early as 1917, *Current Opinion* published "An Indictment of Women's Magazines Edited by Men," which sounded one of the enduring critiques of the magazines: that male control of periodicals intended for women readers was just one more example of men exerting authority over women's lives. The apotheosis of this kind of critique was the 1970 takeover of the *Ladies' Home Journal* offices by representatives of feminist groups demanding, among other things, the ouster of editor John Mack Carter. Thirty-five years before Friedan's book was published, a writer for *Century* magazine expressed similar sentiments in "Woman's Place Is in the Home, So at Least Ten Million Readers Are Urged to Believe." Still others criticized the magazines for being repetitious, for condescending to readers, for assuming that women were responsible for correcting all flaws in both household and marriage, or for creating within their pages worlds far removed from the realities of women's lives. One of the most interesting critiques (not reprinted here because of its length) was launched in the October 1957 issue of *Playboy* magazine. In "The Pious Pornographers," Ivor Williams takes the women's magazines to task for, in his view, pretending to be wholesome guides to household management while offering—in columns by doctors, articles on marital harmony, and fiction—investigations of sexuality not unlike those found in *Playboy* itself.

The articles included in this chapter were originally published between 1946 and 1960 and thus represent voices in opposition to the content of the rest of the selections in this volume. Some were written by authors such as Mary McCarthy and Joan Didion, who were beginning careers as major observers and critics of American culture. Not surprisingly, the periodicals in which such critiques were published had audiences quite different from those of the women's magazines. Most, such as the *Atlantic Monthly* and the *New Republic*, were designed for well-educated, primarily politically liberal readers; others, notably the *Reporter*, commented on the publishing world for an audience largely composed of journalists and other writers. For both groups of readers, the women's magazines were clearly a phenomenon to be reckoned with.

49

ELIZABETH BANCROFT SCHLESINGER[1]

"The Women's Magazines"

New Republic, March 1946

Women's journals, judged by their tremendous circulations, are inexhaustible gold mines for publishers and for the advertisers who use them to compete for the housewife's dollar. They also show what those in control decide should be woman's proper interests. By the same token, their success furnishes a clue to feminine gullibility and literary taste. One might think these periodicals old-fashioned in their devotion to what was once called "woman's sphere," except for the fact that one of them boasts of "the largest circulation of any magazine, given it exclusively by women." That they are popular no one can deny, and therefore one's surprise is all the greater at the triviality of their contents. A study in the *New Republic* in 1933 of the five most widely read publications at that time—the *Ladies' Home Journal*, the *Woman's Home Companion*, *McCall's*, *Pictorial Review* and the *Delineator*—revealed that the old pattern established by *Godey's*, *Graham's* and *Peterson's* a hundred years before was still intact. Their

---

[1]Elizabeth Bancroft Schlesinger (1886–1977), the wife and mother of the historians Arthur Schlesinger and Arthur Schlesinger Jr., was a pioneer in the field of women's history and a lifelong political activist. While in her twenties, she participated in marches advocating woman suffrage, and in her eighties she joined her granddaughter in protests against the Vietnam War.

Elizabeth Bancroft Schlesinger, "The Women's Magazines," *New Republic*, March 11, 1946, 345.

stock in trade was the smartest fashions, new recipes prepared and served in the latest wares of their advertisers, and the cult of making their readers beautiful.

Since that survey was made, twelve of the most eventful years in the lives of women, as well as men, have passed. It is worth while to see whether this experience has affected the content of the monthlies now popular. To find the answer, the *Ladies' Home Journal,* the *Woman's Home Companion* and *McCall's* have been examined over a period of six months (October, 1944, through March, 1945), together with a striking new rival, *Woman's Day,* which sells for two cents over the counter of the Atlantic and Pacific stores. These four, less expensive and less sophisticated than some of the more glamorous and pretentious feminine periodicals of our time, are read by the greatest number of women in the world.

Dumbarton Oaks, Bretton Woods and Yalta[2] were in everyone's mind during this time. There was also a presidential campaign, and the OPA[3] was trying to explain shortages, the advantages of rationing and how to prevent inflation. Moreover, postwar economic and social problems were struggling for consideration and racial clashes were disturbing many communities. These were just a few of the important matters before the nation. Yet even the casual reader would be impressed with the solicitude which the editors displayed in protecting their subscribers from a knowledge of these subjects.

War and its problems lived with us day and night. Therefore it comes as a shock to discover that none of these women's magazines even mentioned Dumbarton Oaks, Bretton Woods or Yalta, or made any reference to the part women could play in bringing about international cooperation for future peace. This does not mean that there were no articles touching on war. On the contrary there were many. Varying in quality and length, they for the most part played up the emotional impact of the conflict on individuals. Only *McCall's* discussed postwar GI education, although thousands of families are vitally interested in the government's efforts to help veterans obtain further training.

Though the most crucial national election of our lifetime occurred in November, just two of the monthlies gave it anything like adequate treatment. *McCall's* had a major article, "Womanpower in the Election," by Charles and Mary Beard, and in its September issue *Woman's Day* featured the event in a piece by Samuel Grafton, entitled "Hey, You Have a Date in November." The *Woman's Home Companion* contented itself

---

[2]Sites in Washington, D.C., New Hampshire, and the Soviet Union, respectively, where major conferences were held to establish the terms of international postwar cooperation.

[3]The Office of Price Administration (OPA) was responsible for controlling the distribution of food during World War II by setting prices and rationing some foods.

with parallel accounts on a single page, in which two women representing the opposing parties gave reasons for their choices. The *Ladies' Home Journal* preferred to treat the crisis in terms of personalities—"First Ladies on Parade," "Meet the Roosevelts" and "Meet the Deweys"—as if America were choosing the nicest man and the best housekeeper.

To turn to other wartime domestic interests, the OPA, despite its valiant efforts to keep prices down and the wolf of inflation from the door, rated only a short editorial and a gossipy account of rationing experiences in the *Ladies' Home Journal,* and was ignored by the other three. Black markets did not exist so far as these periodicals were concerned—a strange commentary on publications which professed to find in food one of their chief interests. As for better racial understanding, a complete blackout prevailed except for a nationalities series in *Woman's Day,* which, however, did not include the Negro, Jewish or Nisei[4] minority groups.

The question naturally arises as to just what the editors and publishers thought American women should be reading during those perilous months. The content of all the magazines was much the same. Food, fashions, beauty hints, home building and decoration, child care, reviews of books, movies and the radio were regular departments. Special features of the *Ladies' Home Journal* included monthly columns by Mrs. Roosevelt and Dorothy Thompson. While it was a pleasure to read Mrs. Roosevelt's honest and wise observations, Miss Thompson, pontificating on miscellaneous subjects, muffed the opportunity to interest her sex in working for peace, fighting racial discrimination or stamping out the black market. The *Woman's Home Companion* from time to time polled its subscribers on different subjects, such as whether husbands should control the family finances, what form war memorials should take and whether it was cricket to further the belief in Santa Claus, all perfectly harmless and not stirring too many brain waves. To its credit, however, it also posed the question of peacetime conscription. In addition, this magazine is to be commended for its regular reports from war correspondents. *McCall's* had a "Washington Newsletter," mostly taken up with current shortages, and a preview of the postwar world in terms of housekeeping. "Neighbors on the Homefront," with its genuine flavor of helpful friendliness, consisted of letters from the readers of *Woman's Day.*

Besides, each periodical gave attention to a variety of subjects, difficult to classify, but evidently aimed at the diversified, if not always significant, interests of the readers. Of *McCall's* eleven pieces, four dealt with

---

[4]Nisei are Japanese Americans born to Japanese parents in the United States.

child care; three with the effect of war on the individual; and one each with disease, the handicapped and a visit to a private home. Child care was the subject of 12 of the 44 articles of the *Woman's Home Companion;* home architecture, six; style designing, five; while two were devoted to treatment of the returning veteran, and two others to physical ailments. The remainder chatted about a school for marriage, the children of Hollywood stars, "five hours to solo" and similar matters. In the 43 articles of the *Ladies' Home Journal,* children rated 10; emotional aspects of the war, six; disease and visits to private homes, four each; morals, three; and education, one.

Woman's Day, the homespun member of this glittering company, was the only one to compliment its readers by discussing important public issues in a significant way. It presented arguments for and against compulsory peacetime military service, appraised prewar college education and stressed women's civic duties. Also among its 31 articles were seven on care of children, three on English literature and five on the serviceman. Though many of the other pieces treated trivial subjects, the writing throughout was marked by a certain down-to-earth quality that set it apart from its rivals. Moreover, its reviews of movies and the radio were the most discriminating of the group.

The inevitable fiction—boy-and-girl tales, generally with happy endings—consumed endless space. Attractively colored illustrations, portraying pretty, smartly dressed women courted by handsome men (usually officers), brought a gay, exciting world into the humdrum existence of many readers.

In the case of all the periodicals, advertising occupied at least half the space, providing an explanation for most of the vagaries of editorial direction. Fingering through the pages of temptingly colored foods recalled a world far removed from ration books. These and other advertisements made it difficult to believe that countless young men were dying on the battlefield. Oddly enough, in the midst of this plentiful display, the War Advertising Council grimly warned, "We must head off inflation now and the best way to do it is to save your money. Buy only what you need. Buy and hold all the War Bonds you can afford."

If this half-year period is typical, such is the content of the magazines which millions of women devour each month. The Lynds in their two books on Middletown[5] say that publications like the *New Republic,* the

---

[5]Robert Lynd (1892–1970) and Helen Lynd (1896–1982) were sociologists who described typical middle-class America in two studies (1929 and 1937) of a midwestern town they called "Middletown."

Nation, Time, Harper's and the *Atlantic* had a combined circulation of only a few hundreds, while the *Ladies' Home Journal, McCall's* and the *Woman's Home Companion* counted their subscribers by the thousands. To put it differently, in many Middletown homes the women's journals were the only ones taken. A similar situation may be assumed in hundreds of other towns and cities throughout the United States. Nor is there any reason to believe that women supplement their information from the daily press. A recent study of what they look at on the front pages of the newspapers shows that twice as many read the local items as read important domestic and foreign news.

The *Ladies' Home Journal* has a circulation of 4,090,659; the *Woman's Home Companion*, 3,586,231; *Woman's Day*, 2,623,202; and *McCall's*, 3,523,350. Such a reading public would give pause to publishers who felt a sense of social responsibility and were able to shed their advertiser and reader fears. These periodicals could easily sell the OPA and the fight against inflation to their readers. They could weld the housewives of the nation into a force that could have destroyed the evil black markets. By championing an Americanism of equal rights and opportunities for all, they could do much to stem the spread of racial misunderstandings. Moreover, they could lead the women of the nation in a crusade to back up international cooperation for world security. All these subjects indubitably fall within the province of the home; indeed, they are the foundation without which the home is endangered.

The modern world demands that more publishers and editors include, along with the traditional interests of woman, her responsibility as a citizen. Such men have not been unknown in the past. Cyrus H. K. Curtis and Edward Bok[6] fought fearlessly for causes in which they believed, even though many times it cost them dearly in circulation and advertising. "Never underestimate the power of a woman," is the slogan of one of these monthlies. It is ironical that the only ones to do so are the magazines themselves.

---

[6]Cyrus H. K. Curtis founded *Ladies' Home Journal* in 1883. Edward Bok edited the magazine from 1889 to 1919.

50

ANN GRIFFITH

"The Magazines Women Read"

American Mercury, March 1949

Women read a great variety of magazines, on a great variety of subjects, and with great variation in their intellectual levels. Some women read *Breezy Love Stories,* and some—not quite so many, perhaps—read the *Harvard Business Review.* But despite the enormous diversity in women's reading habits, it seems to be a fact that the magazines designed *exclusively* for women are all pretty much alike. They are all slick, illustrated, middlebrow and prosperous; and they are all, or nearly all, composed of the same four ingredients: (1) advertisements, (2) social and household hints, (3) "serious" articles, and (4) fiction. The delicate blending of these elements (sometimes it is hard to tell which is which) results in a product that tells us a lot about the modern woman. It would perhaps be overstating the case to say that the women's magazines accurately reflect the spiritual qualities of American womanhood; but they do, certainly, let us in on a few of the problems involved in being an American woman. Take, for example, the problem of cleanliness, which in the women's magazines is quite a problem.

The standard of cleanliness that is set before today's woman is impossibly high. If she reads her ads diligently she must buy new products, or keep on buying old ones, in a hopeless quest for a perfection that eludes her like the Holy Grail. There is no generally agreed upon definition of "clean," and since the term is kept vague and relative she can never be sure that she has achieved it. Suppose she washes her much-maligned hair with a liquid shampoo which claims to be the cleanest shampoo in the world (to cite one of the milder claims). Can she relax now, confident that she has achieved the ultimate in hair-washing? Not if, while her hair is drying, she idly leafs through one of her magazines. The soap-shampoo makers and the creme-shampoo makers will fall upon her with cries of "I can do it better," and will ask her rudely if she has been using one of those *liquid* shampoos that leaves such an ugly, dull film. The three-way struggle between soap-, creme- and liquid-shampoos is so bitter and

---

Ann Griffith, "The Magazines Women Read," *American Mercury*, March 1949, 273.

intense, and the disasters incurred by using any one, according to the other two, so frightening, that it is hard to imagine how any woman can screw up her courage to the point of making a choice. The same difficulty prevails in every area of a woman's life that is subject to scrubbing. No matter what she uses for the floors, for her face, for her husband's shirts, for under the arms, for the sheets, the windows, the top of the stove, or the dishes—a rival product is always needling her to make things bright*er*, clean*er*, fresh*er*, and better smelling. Soap powder ads are pitched on a hysterical level. It used to be that getting rid of tattle-tale gray was enough, but not any more. In a surprising confession of past imperfection, almost every soap company is shouting that *now* it's *really* producing a good soap. "War-time research" has somehow uncovered a number of new ingredients, and if you thought the old formula got your washing clean you're living in a fool's paradise. It is not even enough, any more, for your clothes to be "white-like-new" or "bright-like-new." The latest developments in the laboratory, now incorporated into your package of soap powder, make your things whit*er* and bright*er* than brand-new! There is no end in sight, no hint that there is an optimum whiteness to which you can bring your clothes and then relax.

If it is true that women do believe in their magazines, such tactics must produce a permanent state of uncertainty. To believe that you and everything in your domain must be spotless at all times, and yet to have your faith in whatever cleansing agent you have chosen constantly undermined by its rivals can lead only to frustration. Another virtue that is highly esteemed by the ad-men, good cooking, is similarly impossible to achieve. For every ad that tells how to be cleaner, there are two telling how to have better food, and the same confusion exists for the little woman amidst the pots and pans. She can cook things lighter, fluffier, prettier, more nourishing, more exciting, tastier and crunchier, but again there is no absolute.

Cleanliness and competitive cookery are two of the Big Four drives most hammered at in women's magazine ads. The third may be loosely described as more and better romance. Among the many products which promise more romance, or more satisfactory romance, to the purchaser are the following: corn-plaster ("Who wants a wife with *corns?*"), writing paper, cake, beverages, tapioca, elastic stockings, sheets and bobby pins. (For those who are past caring about romance, there is also a kind of bobby pin that boasts simply "a 144% stronger grip.") Deodorants and soaps stand side by side at the top of the romance list. Their nearest competitor is shampoo, which is followed by perfume, shoes, slips, brassieres, cigarettes, and fingernail polish, in that order. These advertisers are more positive than the cooking and scrubbing boys. They often promise that love will follow the use of their products as surely as the night the day, and probably as quickly.

All too infrequently, the hand of the Federal Trade Commission is faintly discernible, as in the case of the extraordinary admission of a certain "Lustre-Creme Dream Girl." Her facial expression is of such unspeakable vacuity that one would think her hair would have to turn to solid gold for any man to consider having her around the house. But no, apparently only a shampoo is needed to bring her Mr. Handsome to heel. " 'Say,' he exclaimed, 'What's happened to your hair? It looks beautiful, and changes your entire appearance.' " And indeed it has—from plain vacuity to advanced idiocy. But then in the final panel, Miss Dream Girl says, " 'A few weeks later he slipped a sparkler on my finger while he tenderly touched my sparkling hair. I don't suppose it always works this way. But Lustre-Creme Shampoo, with its blend of secret ingredients, *plus lanolin, did* bring out my hair's beauty, *did* attract my man.' " [Italics hers.— A. G.] That belatedly inserted element of doubt could have been only at the behest of the FTC.

The last of the Big Four, the biggest and best whip of all, operates in the kitchen, and includes ads for every modern device and potion that gets the little woman out of the kitchen faster. She is loudly advertised as hating it there, and one out of every two ads in the whole fertile field of women's magazines is concerned with helping her to get through her kitchen-work quicker and more easily. Whether the ads really help her to get out of the kitchen is problematical. But if they do, she does not stay out very long. For the articles have a tendency to send her right back to the kitchen again. Articles designed to keep the woman in the kitchen comprise a hazard of a very special sort. One out of every two contains recipes.

Mostly they are not simple recipes. Mostly, in fact, they might be described as constructions. Salads, for instance, require not only a great deal of preparation time but also a working knowledge of geometry. The ingredients are not mixed comfortably together, but ranged stiffly around a platter, a pattern of cubes, strips, rectangles, circles and triangles. It seems to be much more important that things look pretty, in a sterile sort of way, than that they taste good, though taste is admittedly well thought of. Consider the following:

No one minds getting up from the table to serve himself and, really, with food as pretty to look at as this will be, everyone should see it before your carefully arranged platters are disturbed. While self-service is going on you can do some sleight of hand with the used dishes and your guests will return to a table that looks as charming as it did at the beginning of the meal.

Any hostess who had followed that particular article from beginning to end would hardly be up to any "sleight of hand with the used dishes," and if she had not already collapsed before the guests arrived she would certainly stab to death the first one that laid a fork on her "carefully arranged platters." Who wants the work of sixteen hours, by conservative estimate, to be eaten? Listen to what she has gone through to pull off this simple little dinner party for eight people.

It took her a good three quarters of an hour to read through the menu and write out a shopping list. After checking to be sure that she had the required staples—salt, vinegar, mustard, pepper, sugar, paprika, Worcestershire, Tabasco, cloves, bay leaf, cornstarch, flour, cornmeal, baking powder, baking soda and powdered sugar—and finding that she was lucky enough to have all those on hand, her shopping list still ran to two pages. Somehow she had to get the following stupendous aggregation of groceries home to the kitchen before she even started to work: a five-pound chicken, a veal bone with one pound of veal attached, a veal chop, celery, carrots, milk, bread, thirteen eggs, heavy cream, Chinese seasoning powder, gourmet powder, salad oil, an avocado, garlic juice, butter, blue and yellow vegetable coloring, lots of peas, some parsley, a couple of green peppers, onions, chives, half a pound of shrimp, pimientoes, cream cheese, water chestnuts, fish fillets, lime juice, tomatoes, coconut, okra, mayonnaise, chili sauce, beets, radishes, cauliflower, a calf's foot and a calf's tongue, water cress, butter milk, lemons, cherries, broken nut meats, strawberries, fresh peaches and canned pears.

After having somehow found, and paid for, this curious hodgepodge she still had a long road ahead of her before she could make it come out to look like the pictures. She had to whip or beat thirteen different bowlsful of things. Part way through the master plan the author of this recipe had become apologetic about this aspect and said, "There's a lot of talk about beating here—sorry—and I'm not through yet, but you don't have to knock yourself out." Thereafter she became cagey, and instead of directing, "beat egg whites and add" she had phrased it, "add beaten egg whites." Indirection is common in these recipes. Although designed for the woman with no "help," there is the implication that a staff of fairies is at work in such instructions as "combine 6 cups of peaches, peeled and cut into cubes, with 25 maraschino cherries cut into quarters." "Thin sandwiches should be served with this course" is a happy circumlocution for "make thin sandwiches."

In a few articles there is an Alice-in-Wonderlandish attempt to tackle social problems. The pattern which emerges is of good advice and expensive art work lavished on trivial matters; and of inadequate, unillustrated advice allotted to real problems. Thus eight pages of fashions, in color, will precede an uninspired two-page "treatment" of the conditions in mental institutions. However, by giving even this small space to noted authorities who are known to exist in the real world, the illusion is created that from time to time these magazines do address themselves to reality.

But it is a very circumscribed reality, and on close inspection the illusion becomes insubstantial and all but dissolves. The controversial topics of today are either avoided entirely or presented with so many sides and qualifications and extenuating circumstances that the problem itself tends to sink out of sight. Of four sample articles on marriage, one argued that if your parents' marriage was happy, yours will have a better chance of succeeding; the second held that you should work hard to make your marriage happy so that your children will be happy; the third pointed out that you can expect a change in tempo when the honeymoon is over and you settle down; the fourth advised its readers to learn how to cook and keep house before getting married. Nearly every subject touched upon is drowned in a similar sea of platitudes.

Uncompromising, hard-hitting, straight-down-the-line articles do appear—against tooth-decay or men whistling at women, or in favor of kittens or neighborliness. Like a dog on a leash or a horse in a ring, the reader's mind is let out for a little fresh air, but it is a proscribed bit of exercise, and when it is over she is led gently back into the house, rather than stimulated further.

But if it is true, as advertised, that these *are* the magazines women believe in, then surely the fiction must be the hardest-to-accept of all. Out of a hundred stories, perhaps ten will be concerned with a genuine, recognizable problem. (Which is not to say that the problem will be solved by a genuine solution.)

One dreary couple, for instance, moves to New York after the war and runs spang into the housing problem. Months of apartment-hunting produce no place to live, only disillusionment. Jane would come back to their hotel room at night "exhausted, depressed, sick with the memory of shabby rooms already taken, sinister superintendents, bitter, fighting applicants. Once she witnessed a hair-pulling match between two well-dressed girls who each claimed to be first in line, and she went away trembling, her heart heavy, wondering what had happened to all the decency in the world, the ready laughter and clean eyes of the people she had known before the war." Finally she finds a cold-water flat, but being a loyal reader of the women's magazines, she is miserable here because she can't get it clean enough. Everything, in fact, is too dirty for her—the children on the block, the janitor, the streets, even her dear husband with "clean, square hands," she is wont to muse on, has "the city's grime on the turned-back cuffs of his shirt, and at his collar." She becomes more desperate as the months drone on. All her home-making instincts are thwarted by the limitations of their flat. Quarrels increase, tempers fray. One can follow Ken and Jane this far with some belief, if total lack of interest, because one knows that a housing shortage exists and that a cold-water flat can be hard on the nerves. It is with a heart as heavy as Jane's, however, that one reads the resolution. Comes the Big Quarrel over a burned steak. Bitter words, including the ever popular, "I hate you, I hate you!" Tears. Ken storms out, but is back in a gratifyingly short time. All words are retracted and we are served the following bit of Pablum[1] for a solution. "What did it matter where they lived, as long as they were together? What did it matter about the world if they had each other?"

Such stories, played out against a recognizable backdrop, are rare. The vast majority take place in a never-never land inhabited by disembodied spirits completely free of entangling environments. There is a constant dashing around from Bermuda to Paris to Hawaii to Mexico by people with no visible means of support. The traffic between New York and Reno is exceptionally heavy, and the travelers amazingly light-hearted. There are no time-tables, reservations, or lawyers' fees. Practically everybody lives in New York, Florida or San Francisco.

It takes a shadowy character to survive in a shadowy world. The ability to pull up stakes and dash off wherever love demands is a requisite. Few people have jobs, and almost nobody works from nine to five. Those who do work seem to hold vague positions in advertising agencies or brokerage firms which allow them an unlimited amount of time off. This is a good thing, for husbands are required to spend a great deal of time at home, family life being a series of major crises in which everybody participates. The single man must be away from the office even more than the husband. One indefatigable suitor chased his heroine back and forth across the continent four times before he was able to land her. Unlike Ken and Jane, most of the people in this fiction are free as air, with no worries about jobs, delin-

---

[1] A trade name for an infant cereal similar to oatmeal.

quent children, money, housing, the high cost of living, elections, the state of the world, or any of the problems that dog us mortals on the outside. The contrasts between the people inside the women's magazines and the people the reader must encounter in her daily life are endless. The heroes and heroines are an incredible synthesis of the good, the true and the beautiful, actuated always by the highest motives, as gamely virtuous as Little Orphan Annie. The pre-occupation with cleanliness is again apparent. They are clean morally, and, above all, physically. "Hair nice and clean and red," "clean-jawed," "a good clean profile," "a good clean brow," "a clean, rugged face"—these and a hundred variants are in constant use. All skin is clear and unblemished (one girl was even noted as having "an unblemished foot"), hair is shiny and neat, clothes are pressed, nobody perspires, and the inner man is pure as the driven snow.

Husbands are the most shadowy figures of all in this shadowy world. They seem to have no separate identity, but exist only as projections of their wives. One story went so far as to demonstrate that the hen-pecked man really prefers it that way. The mouse in this case was roused to action when he heard that he'd become the laughing stock of the town. The news had got around that he hadn't been out at night, or taken a drink, or smoked, for ten years. Angrily he announced that he behaved in this fine fashion because he wanted to. (True, his wife had said ten years ago that she "didn't like the smell of smoke.") But he'd show them, nobody was going to say *he* was hen-pecked. Down he goes to the cellar and brings back half a bottle of whisky and half a pack of cigarettes that he'd left there ten years ago, only to discover that *he doesn't really want them,* just as he thought. So he slams on his hat and rushes out of the house— he's still going to show them—but he doesn't know where to go. The finale of this housewife's fantasy finds him, half an hour later, returning to his home and wife, not caring any more if the neighbors do think him hen-pecked, wishing only to be attached to her apron-strings again.

The plots which are acted out by these unrecognizable characters amidst correspondingly unrecognizable settings are not very great variations of the man-hunt. The pursuit of the male, or more specifically the bringing of the male to heel, is the burden of approximately 85 per cent of women's-magazine fiction. The usual plot is almost too banal to bear recounting, being the hoary business of bringing two people together, introducing some barrier to love, demolishing it, and marrying them off. Or, if they are already married, introducing some threat, demolishing it, and leaving them marrieder than ever.

These basic themes having been done an incalculable number of times, a good variation is hard to find, and most of the attempts are pitiable. One man about to marry a girl becomes annoyed because she isn't neat and she drops things around and bangs into things and even sits on the floor sometimes. While he's thinking of calling the whole thing off, she recites a bit of poetry she learned at school. So impressed is he at her knowing a poem that:

> He could not speak for a moment. Something was happening to him deep inside. Like being born. As hard, as fierce, and as anguished.
> He was looking deep into those limpid eyes that missed so much, but that saw and knew what they had to. Young as the morning and ageless in wisdom, those eyes. Her sweetness and sloth, her blindness and understanding—he knew it now—had finally mixed and merged into the sure integrity of the well-beloved.
> He buried his face in her hair; she held him tight. Her arms had gone around his waist, to rest him, to keep him.
> *Darling, my darling, it's going to be all right!* her body said.
> With one hand, he held her away from him; with one hand he touched her eyes, her cheek, and her hair.
> "Please marry me very soon," he said.

One of the reasons the variations are so weak is that the strictest morality must always be observed. A typical hero was "shocked" when a girl he had taken out six times allowed him to kiss her. A more daring story sends a couple who have known each other eight months off for a week end together. Before they reach the first stop she gets cold feet and demands to be taken home. For three months he stays away and she mopes around the house, sure she has lost him, and thinking well, maybe she should have, but then oh no, *no*, better this than that. Virtue is at length triumphantly rewarded and the male is forced to eat humble pie. He does love her, after all; he can't stay away any longer; and finally he appears at the door with a bona fide offer of marriage.

It is time to ask point-blank why women swallow this sort of thing hook, line and sinker?—if they do. What do they get? What is the reward for the tortuous mental gymnastics that women must have to employ to keep their faith in as complicated and contradictory a credo as is handed down to them by their magazines?

A suggested answer is that if they can believe everything they read in them, they can save themselves an endless amount of worrying. A definite aura of "God's in his heaven, all's right with the world" emanates from the great editorial bosom of these magazines. "There, there, little women, leave all your thinking to us" is about what these magazines tell their readers. To be advised on everything, from how much to nip in her waist this month to the proper attitude toward atomic energy is certainly easier for her than thinking for herself. To be assured that she need not bother her pretty head about atomic energy is perhaps reward enough in itself.

I dare say you are neither excited nor bewildered by the fact that men can burn coal and make steam, and with the steam turn generators that produce electric current. The chances are that your only interest in this phenomenon lies in the result of it. When you flip a switch your percolator works, or your lights go on, or your vacuum cleaner starts sucking up dirt. You do not have to know anything whatever about ohms or kilowatts or transformers in order to make use of the energy that comes through your meter in the basement. Atomic energy is the same thing. It is electricity, nothing more. The only difference is in the fuel that is used to produce the electricity. Coal produces electricity because it burns and boils water to make steam. Uranium produces electricity because its atoms can be split apart. This splitting process releases huge amounts of heat. This heat can be used to boil water and make steam, which in turn can be used to make electricity. That is all there is to atomic energy. We have a new fuel.

For her, that's all there need be to it. Atomic energy is just a new means of powering her vacuum—to get things *cleaner.*

51

MARGHANITA LASKI[1]

"What Every Woman Knows by Now"

Atlantic Monthly, May 1950

It is as much a source of amazement as of income to me that readers of the women's magazines have such an insatiable thirst for reading the same information over and over again, despite the fact that any one year's reading must inevitably give enough information about the technique of

---

[1]Marghanita Laski, British journalist, broadcaster, critic, and author, wrote in several different genres. In addition to satiric novels such as *Toasted English* (Boston: Houghton Mifflin, 1949) and suspense novels such as *The Victorian Chaise-Longue* (Boston: Houghton Mifflin, 1954), she wrote books for children, literary criticism, and biography, including *Jane Austen and Her World* (rev. ed., New York: Viking, 1975) and *George Eliot and Her World* (London: Thames and Hudson, 1973).

Marghanita Laski, "What Every Woman Knows by Now," *Atlantic Monthly,* May 1950, 90.

being a woman to see one through a lifetime. I have, then, no fear of spoiling the market, either for myself or others. Every subject in this symposium, given a snappy title and an angle that appeals to the editor, will still be worth a substantial fee.

ACCESSORIES

The simplest are in the best taste.
Men like women to be in the best taste.

BROKEN HEARTS

Find a new interest.
Time cures all.
Men don't like women to ring them up.

CARE OF FACE

Remove old make-up with cream (dry skins), lotion (oily skins), or superfatted soap (if you must).
Then dab face with an astringent lotion.
Then pat in nourishing cream.
Blackheads are frequently due to internal causes. Drink lots of water.
Men are repelled by pimples.

CHARM

Charm is an indefinable quality.
Men like it.

CLOTHES

Choose the clothes that suit you.
You can be perfectly dressed at every income level.
Little touches of white must be immaculate.
Diagonal stripes are slimming.
Invest your all in one good little black dress (or tweed suit).
Don't go in for clutter *but* have lots of bits and pieces that will make one outfit do the work of ten.
Men like black satin, well-cut tweeds, floating tulle, utter simplicity, and don't notice what you wear anyway.

CULTURE

Read good books sometimes.
Men don't like cultured women much.

FIGURE

Figure deficiencies are frequently glandular. Consult your doctor.
A good corset can correct many figure faults. Have it fitted by an expert.
Good exercises can correct bad figures. Here are some.
Men like good figures.

FURS

If you can't afford good furs don't have any, *but* there are some awfully good cheap ones in the shops.
Men are impressed by mink—but then, so are you.

HAIR ON THE HEAD

The condition of the hair reflects the general health.
Massage with the finger tips stimulates the scalp.
Brush fifty times a day and wash at least every fortnight.
Choose the hair style that suits you *and* don't get into a rut.
No moral opprobrium is attached to dye.
Men love those gleaming tresses.

HAIR, SUPERFLUOUS

In the armpits remove by depilatory.
On the legs remove by depilatory, wax, sandpaper, or razor; the last will coarsen the new growth.
On the face remove by wax (will weaken growth) or by electrolysis (will kill it).
If the growth is slight, bleach with peroxide-and-ammonia.
Men notice superfluous hair.

HANDS

Before doing rough work smooth a protective cream over your hands.
After washing, smooth a creamy lotion over your hands.
Make your hands flexible by shaking in one way or another.
File your nails to the shape that suits you.
Press back your cuticles after you've had a bath.
Chipped polish looks slovenly.
Men abhor scarlet talons.

JEWELRY

One big good piece is better than a lot of little cheap trinkets.
One big cheap piece is better than a lot of little good ones.

In fact, One Big Piece is Best.
Men are better if they like jewelry.

MAKE-UP

Smooth on foundation cream or lotion, not forgetting neck.
Add rouge where it improves the natural shape of your face.
Add discreet eye-shadow and mascara on the upper lids only.
Paint outline of lips with a brush, fill in with lipstick, blot on a tissue, powder, and add more lipstick.
Press in powder over face and neck; remove surplus.
Men don't like women to be obviously made up.

MANNERS

Be sweet to old people.
Be kind to his mother.
Be nice to other girls—they have brothers.
Don't comb your hair or clean your nails in public.
Don't order direct from the waiter.
Don't swear or drink too much.
Men hate red marks on coffee cups.

MARRIAGE

Enter it joyously and proudly.
Remember you've got to take as well as give.
There are all sorts of compensations.
Men should be *encouraged* to wash up.

PERFUME

Choose the perfume that suits you.
Spray it onto your body but never onto your clothes.
Test new perfume by trying a drop on the back of your hand.
Have different perfumes for different moods *or* make one perfume distinctively YOU.
Men are enraptured by perfume.

SHOPPING

Either go with an open mind *or* with a rigidly-to-be-adhered-to list.
Either enlist the help of the shop assistant *or* don't let her make up your mind for you.
Either men like shopping *or*—more usually—they don't.

SHYNESS

Prepare a few conversational remarks to break the ice.
Try to put the other person at *his* ease.
Instruct yourself in current affairs.
Join a club.
Men like a woman to be a good listener.

SPECIAL OCCASIONS

Cream hands thickly and sleep in gloves the night before.
Try to fit in a facial and a hair-do.
Rest for an hour with your feet up and pads over your eyes.
Make up extra specially carefully.
Oh, men, men, men.

TOP SECRET

Consult your doctor.
Send us a stamped addressed envelope.
Men are beasts.

52

MARY McCARTHY[1]

"Up the Ladder from Charm *to* Vogue*"*

Reporter, July 1950

"Will you wear a star in your hair at night . . . or a little embroidered black veiling hat? . . . Will you wear a close little choker of pearls or a medal on a long narrow velvet ribbon? . . . Will you serve a lunch, in the garden, of *prosciutto* and melon and a wonderful green salad . . . or sit in the St. Regis's pale-pink roof and eat *truite bleue?*"

It is the "Make Up Your Mind" issue: *Vogue*'s editresses are gently pressing the reader, in the vise of these velvet alternatives, to choose the looks that will "add up" to *her* look, the thing that is hers alone. "Will you make the point of your room a witty screen of drawings cadged from your artist friends . . . or spend your all on a magnificent carpet of flowers that decorates and almost furnishes the room itself?"

Twenty years ago, when *Vogue* was on the sewing room table of nearly every respectable upper-middle-class American house, these sapphic overtures to the subscriber, this flattery, these shared securities of *prosciutto* and *wonderful* and *witty* had no place in fashion's realm. *Vogue,* in those days before *Mademoiselle* and *Glamour* and *Charm* and *Seventeen,* was an almost forbidding monitor enforcing the discipline of Paris. An iron conception of the mode governed its semimonthly rulings. Fashion was distinguished from dress; the woman of fashion, by definition, was a woman of a certain income whose clothes spoke the idiom of luxury and bon ton; there was no compromise with this principle. Furs, jewels, sumptuous materials, fine leathers, line, cut, atelier workmanship, were the very fabric of fashion; taste, indeed, was insisted on, but taste with-

[1]Mary McCarthy (1912–89), novelist and literary and political critic, began her career with the novel *The Company She Keeps* (New York: Harcourt, Brace, 1942) and gained wide public attention with her 1963 novel *The Group* (New York: Harcourt, Brace & World). Her several autobiographical volumes include *Memories of a Catholic Girlhood* (New York: Harcourt, Brace, 1957), *How I Grew* (San Diego: Harcourt, Brace, Jovanovich, 1987), and *Intellectual Memoirs: New York, 1936–1938* (New York: Harcourt, Brace, Jovanovich, 1992). Some of her essays on social issues are collected in *On the Contrary* (New York: Farrar, Straus, and Cudahy, 1961), and in the last years of her life she was a regular contributor to the *New York Review of Books.*

Mary McCarthy, "Up the Ladder from *Charm* to *Vogue," Reporter,* July 18, 1950, 36.

out money had a starved and middle-class pathos. The tastefully dressed little woman could not be a woman of style.

To its provincial subscribers *Vogue* of that epoch was cruel, rather in the manner of an upper servant. Its sole concession to their existence was a pattern department. *Vogue's Designs for Dressmaking,* the relic of an earlier period when no American woman bought clothes in a shop. And these patterns, hard to cut out as they were, fraught with tears for the amateur, who was safer with the trusty Butterick, had an economical and serviceable look that set them off from the designer fashions: Even in the sketches they resembled maternity dresses.

As for the columns of etiquette, the bridal advice, the social notes from New York, Philadelphia, San Francisco—all these pointedly declined acquaintance with the woman-from-outside who was probably their principal devotee. Yet the magazine was read eagerly and without affront. Southern women, Western women with moderate incomes pored over it to pick up "hints," carried it with them to the family dressmaker, copied, approximated, with a sense, almost, of pilferage. The fashion ideas they lifted made the pulse of the Singer[2] race in nervous daring and defiance (What would *Vogue* say if it knew?).

This paradoxical relation between magazine and audience had a certain moral beauty, at least on the subscribers' side—the beauty of unrequited love and of unflinching service to an ideal that is arbitrary, unsociable, and rejecting, like Kierkegaard's God and Kafka's Castle.[3] Lanvin, Paquin, Chanel, Worth, Vionnet, Alix[4]—these stars of the Paris firmament were worshiped and charted in their courses by reverent masses of feminine astronomers who would come no closer to their deities than to copy, say, the characteristic fagoting that Vionnet used in her dress yoke or treasure a bottle of Chanel's Number Two on the bureau, next to father's or husband's photograph.

Like its competitor, *Harper's Bazaar,* and following the French dressmaking tradition, *Vogue* centered about the mature woman, the *femme du monde,* the sophisticated young matron with her clubs, her charities, and her card-case. The jewels, the rich fabrics, the furs and plumes, the exquisite corseting, the jabots and fringe, implied a sexual as well as a material opulence, something preening, flavorsome, and well satisfied.

---

[2]Sewing machine manufactured by the Singer Company. (Although there were other brands, the name Singer became synonymous with "sewing machine.")

[3]Søren Kierkegaard (1813–55), Danish existential philosopher; Franz Kafka (1883–1924), Austrian writer whose unfinished novel *The Castle* (1926) sets forth a metaphor for an unyielding, incomprehensible God.

[4]Trend-setting Paris designers of high fashion.

For the *jeune fille* (so defined) there was a page or two of party frocks, cut usually along princess lines, in pastel taffetas, with round necks. In this Racinean[5] world, where stepmother Phèdre and grandmother Athalie queened it, the actual habits of the American young girl, who smoked and wore lipstick, were excised from consideration. Reality was inferior to style. Covertly, the assumptions of this period remain in force. Despite social change, fashion is still luxurious. It is possible to dress prettily on a working girl's or business wife's income, but to dress handsomely is another matter, requiring, as before, time, care, and money. Fashion is a craft, not an industrial conception, exemplifying to perfection the labor theory of value. The toil of many hands is the sine qua non of fashion. The hand of the weaver, the cutter, the fitter, the needleworker must be seen in the finished product in a hundred little details, and fashion knowledge, professionally, consists in the recognition and appraisal of the *work* that has gone into a costume. In gores and gussets and seams, in the polish of leather and its softness, the signature of painstaking labor must be legible to the discerning, or the woman is not fashionably dressed. The hand-knit sweater is superior to the machine-knit, not because it is more perfect, but on the contrary because its slight imperfections reveal it to be *hand*-knit. The Oriental pearl is preferred to the fine cultured pearl because the marine labor of a dark diver secured it, a prize wrested from the depths, and the woman who wears Oriental pearls believes that they show variations in temperature or that they change color with her skin or get sick when they are put away in the safe—in short, that they are alive, whereas cultured pearls, mass stimulated in mass beds of oysters, are not. This sense of the accrued labor of others as a complement to one's personality, as *tribute* in a double sense, is intrinsic to the fashionable imagination, which desires to *feel* that labor next to its skin, in the hidden stitching of its underwear— hence the passion for handmade lingerie even among women whose outer clothing comes off the budget rack.

In spite of these facts, which are known to most women, if only in the form of a sudden anguish or hopelessness ("Why can't *I* look like that?"), a rhetoric of fashion as democracy, as an inherent right or manufacturer's guarantee, has swept over the style world and created a new fashion public, a new fashion prose, and a whole hierarchy of new fashion

[5]Jean Baptiste Racine (1639–99) was a French dramatist. Two of his plays were *Phèdre* (1677) and *Athalie* (1691).

magazines. *Mademoiselle, Glamour, Charm* —respectively "the magazine for smart young women," "for the girl with the job," "the magazine for the B. G. [Business Girl]," offer to the girl without means, the lonely heart, and the drudge, participation in the events of fashion, a sense of belonging en masse and yet separately, individually, of being designed for, shopped for, read for, predicted for, cherished. The attention and care and consideration lavished on the woman of leisure by lady's maid, coiffeur, *vendeuse*,[6] bootmaker, jeweler, are now at the disposal of the masses through the various Shophounds, Mlles. Wearybones, beauty editors, culture advisers, male and female confidants. The impersonally conceived Well-Dressed Woman of the old *Vogue* ("What the Well-Dressed Woman Will Wear") is tutoyered,[7] so to speak, as *You* ("Will you wear a star in your hair? . . . "): and a tone of mixed homage and familiarity: "For you who are young and pretty," "For you who have more taste than money," gives the pronoun a custom air.

The idea of a custom approach to ready-made, popular-priced merchandise was first developed by *Mademoiselle*, a Street and Smith publication launched during the depression, which differed from *Vogue* and the *Bazaar*, on the one hand, and from *McCall's* and *Pictorial Review*, expressions of the housewife, on the other. Before the depression, there had been, roughly speaking, only three types of women's apparel: the custom dress, the better dress, and the budget or basement dress. Out of the depression came the college shop and out of this the whole institutionalized fiction of the "debutante" shop and the "young-timers'" floor. These departments, which from the very outset were swarming with middle-aged shoppers, introduced a new category of merchandise: the "young" dress, followed by the "young" hat, the "young" shoe, the "young" petticoat, and so on. The "young" dress was a budget dress with status, an ephemeral sort of dress, very often—a dress that excited comment and did not stand up very well. Its popularity proved the existence of a new buying public, of high-school and college girls, secretaries, and office workers, whose dress requirements were very different from those of the busy housewife or matron. What these buyers demanded, for obvious vocational reasons, was not a durable dress or a dress for special occasions, even, but the kind of dress that would provoke compliments from co-workers, fellow students, bosses—a dress that could be discarded after a few months or transformed by accessories into the simulation of a new dress. To this public, with its craving for popularity, its personal-

---

[6]Saleswoman.
[7]Addressed familiarly, that is, with the pronoun *you*.

ity problems, and limited income, *Mademoiselle* addressed itself as "your" magazine, the magazine styled for *you*, individually.

Unlike the older magazines, whose editresses were matrons who wore (and still wear) their hats at their desks as though at a committee meeting at the Colony Club, *Mademoiselle* was staffed by young women of no social pretensions, college graduates and business types, live wires and prom queens, middle-class girls peppy or sultry, fond of fun and phonograph records. Its tone was gamely collegiate, a form of compliment, perhaps, since its average reader, one would have guessed, was either beyond college or below it, a secretary or a high-school student. It printed fiction—generally concerned with the problems of adolescence—job news and hints, beauty advice, and pages of popular-priced fashions photographed in technicolor or Burpee-catalogue hues against glamorous backgrounds. Its models were wind-swept and cute.

Fashion as fun became *Mademoiselle*'s identifying byword, a natural corollary to the youth theme. *Fun* with food, *tricks* with spices, herbal *magic,* Hawaiian pineapple, Hawaiian ham, Hawaiian bathing trunks, Hollywood playclothes, cruise news, casserole cookery, Bar-B-Q sauce reflect the dream mentality of a civilization of office conscripts to whom the day off, the two weeks basking in the sun during February or August, represent not only youth but an effortless, will-less slack season *(slacks, loafers,* hostess *pajamas),* quite different from the dynamic good time of the 1920's.

In the *Mademoiselle* play world, everything is romp-diminutive or make-believe. The beau is a "cute brute," the husband a "sahib," or "himself," or "the little fellow." The ready-mix cake "turns out *terrific."* Zircons are "almost indistinguishable from diamonds." "Little tricks of combination, flavor and garnishment help the bride and enchant the groom . . . who need never know!" Brides wearing thirty-five-dollar dresses are shown being toasted in champagne by ushers in ascots and striped trousers.

Work may be fun also. "I meet headline people on the Hill every day." Husband-and-wife *teams* do "the exciting things" together. And the work-life of a reader-surrogate named Joan, *Mademoiselle's* Everygirl, is to be continually photographed backstage at "exciting" events, "meeting summer halfway on a Caribbean island," meeting Maurice Evans in his dressing room, or gapily watching a chorus rehearsal. The word *meet,* in the sense of "coming into contact with or proximity of," is a denotation of holiday achievement. Resort news is eternal, like hotel-folder sunshine.

The strain of keeping up this bright deception is marked by the grotesquerie of adverbs ("Serve piping hot with a dish of wildly hot mustard nearby"), by the repeated exclamation point, like a jerky, convulsive party smile, and by garish photographic effects. The typical *Mademoiselle* model with her adolescent, adenoidal face, snub nose, low forehead, and perpetually parted lips is immature in an almost painful fashion—on the plane, in the Parisian street, or the tropic hotel she appears out of place and ill at ease, and the photography which strives to "naturalize" her in exotic or expensive surroundings only elevates her further. Against the marble columns or the balustrades, with fishing rod, sailboat, or native basket, she stands in a molar eternity, waving, gesticulating, like the figures in home movies of the vacation trip. ("See, there she is, feeding the pigeons; see, that's Mabel there by the azaleas.")

Another magazine, *Seventeen*, which from its recipes and correspondence column appears to be really directed to teen-agers and their problems, strikes, by contrast with *Mademoiselle*, a grave and decorous note. Poorly gotten out and cheaply written, it has, nevertheless, an authentic small-town air; more than half of its circulation is in towns under twenty-five thousand. It is not, strictly speaking, a fashion magazine (though it carries pages of fashion, gifts, and designs for knitting and dressmaking), but rather a home magazine on the order of *Woman's Home Companion*. How to make things at home, simple dishes to surprise the family with, games to play at parties, nonalcoholic punches for after skating, candies, popcorn balls, how to understand your parents, how to stop a family quarrel, movies of social import, the management of the high-school prom, stories about friendships with boys, crushes on teachers, a department of poems and stories written by teen-agers—all this imparts in a rather homiletic vein the daily lesson of growth and character building.

Pleasures here are wholesome, groupy ("Get your gang together") projects, requiring everybody's cooperation. Thoughtfulness is the motto. The difficulty of being both good and popular, and the tension between the two aims (the great crux of choice for adolescence), are the staple matter of the fiction; every boy hero or girl heroine has a bitter pill to swallow in the ending. The same old-fashioned moral principles are brought to bear on fashion and cooking. The little cook in *Seventeen* is not encouraged, in the *Mademoiselle* style, to think she can make "high drama" out of a Drake's Cake and a pudding mix; she starts her party biscuits or her cake with fresh eggs, fresh butter, and sifted flour. Her first grown-up jewelry is not an "important-looking" chunk of glass but a modest gold safety pin or, if she is lucky and an uncle can give it to her for graduation, a simple gold wrist watch.

And in *Seventeen*, strangely enough, the fashions, while inexpensive, have a more mundane look than *Mademoiselle*'s dresses, which tend to be junky—short-waisted, cute, with too many tucks, pleats, belts, and collars for the money. The *Seventeen* date dress is not very different from the "young fashion" in *Vogue* or *Harper's Bazaar*. It has been chosen to give its wearer a little air of style and maturity, on the same principle that an actor playing a drunk tries, not to stagger, but to walk straight. The artifice of youth in the *Mademoiselle* fashions betrays the very thing it is meant to cover—cheapness—and the little short bobbing jackets and boleros and dirndls become a sort of class uniform of the office worker, an assent to permanent juniority as a form of second-class citizenship, on the drugstore stool.

In the upper fashion world, the notion of fashion as fun acquires a delicate savor. The *amusing,* the *witty,* the *delicious* ("a deliciously oversized stole") evoke a pastoral atmosphere, a Louis Seize[8] scene where the queen is in the dairy and pauperdom is Arcadia. The whim, piquant or costly, defines the personality: Try (*Harper's Bazaar*) having *everything* slip-covered in pale Irish linen, including the typewriter and the bird cage; and "just for the fun of it, black with one white glove." The idea of spending as thrift, lately coined by *Vogue,* implies the pastoral opposite of thrift as the gayest extravagance. "There is the good handbag. The pairs of good shoe. . . . The wealth-to-spare look of rich and lean cloths together." A "timeless" gold cross made from old family stones, and seventy-dollar shoes are proposed under the heading "Economical Extravagances." "And upkeep, extravagantly good, is the ultimate economy. Examples: having your books with fine bindings oiled by an expert every year or having your wooden shoe-trees made to order. . . . And purely for pleasure, flowers, silver, and the price of keeping it polished; an Afghan hound, the collection, from stamps to butterflies, to Coalport cabbages, that you, or me skimp for rather than do without."
The fabrication here of a democratic snobbery, a snobbery for everyone, is *Vogue*'s answer to the tumbrils of Truman. The trend of the times is resolutely reckoned with: Today "the smaller collectors who have only one Giorgione" buy at Knoedler's Gallery, just as Mellon used to do. As John Jacob Astor III[9] said, "A man who has a million dollars is as well off as if he were rich." (What a *delicious* sow's ear, my dear, where did you *get* it?) The *small* collection, the *little* evening imply the intimate and the choice, as well as the tiniest pinch of necessity. *Little* hats, *little*

---

[8]Louis XVI (*Seize* in French) (1754–93), French king.
[9]Andrew Mellon (1855–1937), American financier and art connoisseur who endowed the National Gallery of Art; John Jacob Astor III (1864–1912), American financier who built the Waldorf-Astoria Hotel in New York City and who went down with the *Titanic* in 1912.

furs, *tiny* waists—*Vogue* and the *Bazaar* are wriggling with them; in the old days hats were *small*. And as some images of size contract or cuddle ("Exciting too the tight skull of a hat with no hair showing"; "the sharp, small, polished head"), others stretch to wrap and protect: *enormous, huge, immense*—"a colossal muff," "vast" sleeves; how to have *enormous* eyes. By these semantic devices the reader is made to feel small, frail and valuable. The vocabulary has become extremely tactile and sensuous, the caress of fine fabrics and workmanship being replaced by the caress of prose.

The erotic element always present in fashion, the kiss of loving labor on the body, is now overtly expressed by language. Belts *hug* or *clasp;* necklines *plunge;* jerseys *bind.* The word *exciting* tingles everywhere. "An outrageous amount of S.A."[10] is promised by a new makeup; a bow is "a shameless piece of flattery." A dress is no longer low-cut but *bare.* The diction is full of movement: "hair swept all to one side and just one enormous earring on the bare side." A waist rises from a skirt "like the stem of a flower." Images from sport and machinery (*team, spark*) give this murmurous romanticism a down-to-business, American twang and heighten the kinetic effect. "First a small shopping expedition . . . . Then give your mind a good going-over, stiffen it with some well-starched prose; apply a gloss of poetry, two coats at least."

[10]Sex appeal.

53

KATHARINE M. BYRNE

"Happy Little Wives and Mothers"

America,[1] January 1956

The happy little wife and mother is really busy these days, and she is making my life no easier. You cannot turn many pages of a Catholic magazine without running into the brave and cheerful story of her life. Her days are filled with worthy projects at home and abroad, and the modest recital

[1]*America* is a Catholic weekly published by the Jesuits.

Katharine M. Byrne, "Happy Little Wives and Mothers," *America*, January 28, 1956, 474.

of her successes ("Of course, I can't do very much as I have eight children under six years of age") must have some good purpose in mind. Perhaps she rushes into print as an encouragement to the spiritually-lagging or hollering-at-the-kids type of female. That a quite opposite effect may follow is certainly no fault of hers.

The Life Beautiful

Most of us females of a lower order have a hard enough time learning to live with the lady in the *House Gracious* ads. You know the one. She sits smiling on her sun-drenched patio reading *House Gracious*. In the out-of-doors recreation area, some distance removed, her two roguish children ("We will raise a family, a boy for you, a girl for me") are engaged in constructive and compatible play. Or she may be sitting at a desk in the meal-planning area of her kitchen, her gourmet cookbook at her fingertips, a pink telephone at her elbow. No child has ever hurled a bowl of Pablum at these walls, nor is this gleaming floor ever awash with spilled Wheaties.

Poor banished children of Eve, we look with longing at all the Things which seem to fill her life so nicely. Only after a spiritual struggle which may last for years do we learn to rise above this girl, and to reject her way of life as false, materialistic and secular. Besides, we tell ourselves, she couldn't be that happy. Maybe she has a mean mother-in-law.

But we are faced at times with a different and more difficult problem. There is another Happy Little Wife and Mother who sits on no sun-drenched patio. She lives, usually, in a huge lovable wreck of a house, distracted by few modern conveniences. In some cases her numerous brood may be tucked into a three-room apartment. No matter. Cheerful as a well-worn cliché, she makes out nobly. While you pale at the thought of 48 hours with a non-operating Bendix,[2] she wouldn't mind beating the bluejeans on rocks.

Her children are good. Her curly-headed two-year-old folds dimpled hands in prayer. Yours has just sunk savage teeth into the arm of her little brother, and followed up his screams with a soothing kiss. No such ambivalent behavior ruffles the spiritual calm of her household.

You may think you are doing a fair job in human relations, but your efforts never work out quite as beautifully as hers. If she is good to the little boy nobody in the neighborhood likes, he blossoms under her kindly ministrations, is diverted from his objectionable hobby (stealing small articles from the local dime store), and now writes her grateful letters from a monastery.

[2]A brand of washing machine.

There was a little boy who hung around your swings and sandbox one summer. Nobody knew where he lived, and the other children weren't very kind to him. When you brought out the milk and sandwiches for yard picnics, you used to bring some for him, too. You urged the children to share toys and popsicles with this outlander. One day when it was time to put the rolling stock away, you noticed that one fairly new 24-inch bike was missing. You never saw the little boy again either.

Nothing like this ever happens to the Happy Little Wife and Mother. Hers is a simplicist's world of easy and invariable answers to life's questions, a kind of you-too-can-learn-to-play-the-Hawaiian-guitar or *Readers' Digest* World in which formulas are neat and all the experiments behave as they should.

And yet, you know that life cannot be so simple, even for her. She probably leads the same soul-buffeted life that we do. She may be better at it, but she's human, and I wish she'd break down and admit it. It would be a real comfort to me to hear the H. L. W. and M. admit that once, after three bleak winter weeks of unalleviated pressures, she walked out on her whole family and took a bus ride to the end of the line, alone.

When I was a little girl we had a remarkable neighbor named Mrs. Mulholland. Because she was the oldest person in the community, her birthday was always the occasion for a newspaper interview. When she reached her 100th year the usual questions were asked. But Mrs. Mulholland, God love her, had none of the usual answers. Did she drink? Well, yes, a little. She had had her first cocktail at 95. Wasn't that a bit late in life to start drinking? "Well, no. Before that I just took a straight nip when I needed it." Wasn't it hard for her to raise all those children alone, since her husband had died when she was in her thirties? Well, no, not as hard as you might think. Her husband, though a good man, you understand, had never really been too much help to her. But she had had a bachelor brother, Joe, with a good civil service job with the city, and he had turned over his check for years. Dear Mrs. Mulholland, I salute your honest virtues.

The Human Touch

When a woman whose dieting efforts have largely failed reads a "You, too ... " article by a lady who lost 30 pounds in 50 weeks, she is heartened by the author's rueful admission that once, in the midst of this rigorous regimen, she locked herself in the bathroom and devoured a pound of butter-creams.

In much the same way, perhaps, we would welcome from the Happy

Little Wife and Mother the admission that while the way of life which she chose, and the one which, with God's grace she is trying to live well, is the one she wants, it is nevertheless a somewhat monotonous life. And often very lonely.

And on occasion, as she kisses her immaculate, clean-shaven, white-collared husband goodbye, and turns to face the montage of congealed egg-yolks, unbraided braids, ankle-deep cereal and damp baby which constitute her first order of the day, might she not indulge, even briefly, her Cool Sewer Complex? (This complex was inspired by Ed Young's famous cartoon depicting the fat and harassed wife who greets her Art Carney-type husband[3] with the classic plaint, "Here I am, standing all day over this hot stove, while you're down in that cool sewer.") Or think, even fleetingly, "Lord, life was good in the dime store." Or the Acme Tool and Die Works. Or the dust and dimness of the Modern Language Library stacks.

While I am often plunged into sadness by a comparison of my own inadequacies with the lives led, in print, by all Happy Little Wives and Mothers, I would feel a real spiritual affinity for the woman who will give us groundlings a work-in-progress report of her efforts toward the Good Life, an account written, not from the peak of Everest, but from halfway up, where the going is still rough and the backslides many.

[3]The actor Art Carney played Jackie Gleason's working-class buddy in *The Honeymooners* series on television.

54

JOAN DIDION[1]

"Marriage à la Mode"

National Review, August 1960

I write as one incurably addicted to the women's "service" magazines: *Ladies' Home Journal, McCall's, Good Housekeeping*. As such, I have long been schooled in certain tenets of contemporary life which may well have passed you by.

1) Women's magazines admit two kinds of women: Homemakers and Housewives. "Returning from PTA the other night," cooed a *Journal* Homemaker not long ago, "I suddenly realized that, in a way, our house is a lighthouse—and I am still the keeper's sweetheart . . . Our lives are blessed with such freshening variety!" '*Freshening variety*,' " snarled a *Journal* Housewife a few issues later. "Spare us the mockery of that! Drag out of bed at 6:30 A.M., prepare breakfast, pack lunches, wash eggy dishes, make beds, wash clothes, iron clothes, scrub the floors, and the neverending battle with dust, dust, dust!"

But whether 2) Homemakers or Housewives, women carry on the world's work. That work 2, A) is *never done*.

Although money 3) never brings happiness, electrical appliances often do. "Stop being a patsy," one Mary Haworth reader advises another. "Any family that can afford to eat can afford a dishwasher." Women 4) are frigid. To a recent letter from a lady who claimed not to be, the *Journal* affixed the title "The Other Side of the Coin."

Men 5) are almost as amusing as children, besides being a help around the house, at least when they aren't out Gambling, Letching at Office Parties, or Taking Themselves Seriously.

In any case 6), men can be handled.

---

[1]Joan Didion (b. 1934), novelist and political journalist, worked for *Vogue* and wrote columns for *Esquire* and *New West* early in her career. Her novels include *Run River* (New York: I. Oblensky, 1963), *Play It As It Lays* (New York: Farrar, Straus & Giroux, 1970), and *Democracy* (New York: Simon and Schuster, 1984); and her volumes of political commentary include *Slouching toward Bethlehem* (New York: Farrar, Straus & Giroux, 1968), *The White Album* (New York: Simon and Schuster, 1979), *Salvador* (New York: Simon and Schuster, 1982), and *Miami* (New York: Simon and Schuster, 1987).

Joan Didion, "Marriage à la Mode," *National Review*, August 13, 1960, 90.

This handling, this marital know-how, is the basis of all women's magazine marriage counseling. These days, since *McCall's* has gone all out on Dali[2] illustrations and articles called "I Am a Geisha," and *Good Housekeeping* has missed the boat altogether (covers bearing photographs of small girls in raincoats and the words "Plastics Bring a New Way of Life" do not, I think, compel any but the hopelessly hooked, like me), the *Journal* is more or less single-handedly bearing the torch of marriage counseling. The *Journal* has Goodrich C. Schauffler, M.D. and "Tell Me Doctor" (Intimate Problems). The *Journal* has Clifford R. Adams, Ph.D. and "Making Marriage Work." (The fact that I am unmarried has never deterred me from taking—and, with relentless regularity, failing—the *Journal*'s monthly "Making Marriage Work" test.)

And, above all, the *Journal* has Paul Popenoe, Sc.D. and "Can This Marriage Be Saved?" Dr. Popenoe, a not-quite-professional man who presides over the American Institute of Family Relations out on Sunset Boulevard, has now collected—with the assistance of Dorothy Cameron Disney, a *Journal* lady—twenty of his *Journal* case histories in a single memorable volume: *Can This Marriage Be Saved?* (Macmillan, $4.95). Out there on the Strip, the *specialité de L'Institut* is a rare, heady, enchanting blend of penetrating insights (when "a husband tells his wife that he no longer loves her and is desperately in love with some other woman," counsels Dr. Popenoe, she should "recognize that such a declaration is frequently based on emotion rather than reason"), considered warnings ("At the Institute we are not enthusiastic about office parties") and reliance upon know-how.

The real trick, be not deceived, is this *savoir-faire.* Good Management. Handling. Consider Andrea Weymer, whose husband, Dick, was about to leave her for The Other Woman. Through counseling, Andrea acquired the know-how to fight for her man, a cool battle which involved her losing eight pounds, learning "to choose smarter clothes in more becoming colors," and, I swear to God, joining the League of Women Voters. "A couple of months after joining the League, Andrea surprised her family by besting Dick in a dinner-table argument. She convinced him it was his civic duty to register and go to the polls."

Consider Elise Manning, who faced the same problem. After a chat with Dr. Popenoe, Elise "found time for swimming, hiking, and tennis, and insisted that he [her erring mate] make time to enjoy these recreations with his family . . . She then began to spend more on herself and

---

[2]Salvador Dali (1904–89), Spanish surrealist artist.

the children and gave him the satisfaction of feeling like an extra-good provider . . . Elise is sure there is no danger of his ever becoming interested in another woman. She and the children occupy too much of his time." (And, although Dr. Popenoe leaves the point implicit, his money.) Although you and I might think, *a priori,* that Elise has one or two surprises ahead of her yet, Dr. Popenoe thinks not.

Again, consider Jill Lester, who complained: "Bob is two inches shorter than me, and the two of us look simply ridiculous together. How can a wife be proud of a husband shorter than she is? . . . I don't care too much about sex. In my opinion sex is — is messy. Mamma thinks the same. And you should hear Alice [a sister] and my three aunties on the subject." Watch the subtlety with which the Institute tackled Jill's problem: "We did not argue with Jill. We did talk to her in a casual way about the possible disadvantage of her future without Bob." (Dr. Popenoe means M-O-N-E-Y.) "But she still could not endure the fact that Bob was two inches shorter than she was . . . This difficulty was solved in a very simple way. Bob bought a pair of the so-called 'elevator' shoes and, so far as ordinary observers were concerned, became as tall as Jill."

Dr. Popenoe, however, puts most of his faith in Talking Things Out. Although it has been my bad luck and/or bad management never to have met a man with whom things might be Talked Out, I accept Dr. Popenoe's word that it remains the most reasonable of tactics. "A clever wife," says he (the crucial adjective here, I suppose, is "clever"), "can often get action by approaching the problem of her mother-in-law indirectly. Picking a quiet time, she can say to her husband, 'I have been wondering if you feel I am dominated too much, in some ways, by my mother? Tell me frankly, and you and I together can work out a solution.' "

Although I can scarce believe that this gambit would elicit any response other than a very fishy eye, Dr. Popenoe claims otherwise. "Maybe he will reply with astonishment, 'I never dreamed of such a thing!' Maybe his reply will contain some sharp criticisms. In either event he will probably follow his comments with 'Do you think I am paying too much attention to my own mother?' Then the two can talk the matter out."

Talking Things Out need not stop, according to the Institute, when no immediate Mummy problems are at hand. "In some families," counsels Dr. Popenoe, "it is worthwhile to hold a monthly fact-finding session to clear the air. These couples set aside the 'mensiversary' (not anniversary) of their wedding for the purpose. On the evening of the twenty-third of each month, say, they go out to dinner and after a pleasant meal in a quiet place they bring forth the topics they want to discuss. They keep a list during the month, probably eliminating a few items every time they add new ones. Over the dinner table, calmly and cooperatively (it is to be hoped), they talk out and agree on how to settle the various unsettled matters in their minds."

Calmly and cooperatively (it is to be hoped), *noli me tangere*,[3] Dr. Popenoe.

[3]Touch me not (Latin).

Questions for Consideration

1. What views of woman's "proper role" are reflected in these articles from women's magazines? How are some of these views contradictory?
2. What vision of the American family emerges from these articles? What assumptions do the magazines seem to make about family life?
3. What particular impact do the articles show World War II having on women's lives?
4. What social class differences do you see in articles from different magazines?
5. What do the articles suggest about the importance of consumer goods in American life?
6. Do the visual images in some of the articles and in the advertisements included in the book contribute to your understanding of the period? If so, how?
7. What elements, issues, or arguments in these articles might be explained by the fact that the period during which they were published immediately followed a decade of severe economic depression?
8. What categories and kinds of American women are *absent* from these articles?
9. What issues raised in the articles are still the subject of controversy today?
10. What kinds of criticisms of women's magazines are expressed in the articles in Chapter 6?

Suggestions for Further Reading

Beard, Charles A., and Mary R. Beard. *The Rise of American Civilization*. Vol. 3 of *America in Midpassage*. New York: Macmillan, 1939.

Brown, Bruce W. *Images of Family Life in Magazine Advertising: 1920–1978*. New York: Praeger, 1981.

Brown, John Mason, ed. *The* Ladies' Home Journal *Treasury*. New York: Simon and Schuster, 1956.

Cancian, Francesca M., and Steven L. Gordon. "Changing Emotion Norms in Marriage: Love and Anger in U.S. Women's Magazines since 1900." *Gender and Society* 2, no. 3 (September 1988): 308–42.

Chafe, William H. *The Paradox of Change: American Women in the 20th Century*. New York: Oxford University Press, 1991.

———. *The Unfinished Journey: America Since World War II*. 2nd ed. New York: Oxford University Press, 1991.

Cohn, David L. *Love in America: An Informal Study of Manners and Morals in American Marriage*. New York: Simon and Schuster, 1943.

Coontz, Stephanie. *The Way We Never Were: American Families and the Nostalgia Trap*. New York: Basic Books, 1992.

Covert, Catherine L., and John D. Stevens. *Mass Media between the Wars: Perceptions of Cultural Tension, 1918–1941*. Syracuse: Syracuse University Press, 1984.

Cowan, Ruth Schwartz. *More Work for Mother: The Ironies of Household Technology from the Open Hearth to the Microwave*. New York: Basic Books, 1982.

———. "Two Washes in the Morning and a Bridge Party at Night: The American Housewife between the Wars," *Women's Studies* 3 (1976): 147–72.

Damon-Moore, Helen. *Magazines for the Millions: Gender and Commerce in the* Ladies' Home Journal *and the* Saturday Evening Post, *1880–1910*. Albany: State University of New York Press, 1994.

Elder, Donald, ed. *The Good Housekeeping Treasury*. New York: Simon and Schuster, 1960.

Ehrenreich, Barbara. *The Hearts of Men: American Dreams and the Flight from Commitment*. Garden City, N.Y.: Doubleday, 1983.

———, and Deirdre English. *For Her Own Good: 150 Years of the Experts' Advice to Women*. Garden City, N.Y.: Doubleday, Anchor Press, 1978.

Ferguson, Marjorie. *Forever Feminine: Women's Magazines and the Cult of Femininity*. London: Heinemann, 1983.

Fox, Bonnie J. "Selling the Mechanized Household: 70 Years of Ads in *Ladies' Home Journal*." *Gender and Society* 4, no. 1 (March 1990): 25–40.

Fox, Richard Wightman, and T. J. Jackson Lears, eds. *The Culture of Consumption: Critical Essays in American History, 1880–1980*. New York: Pantheon, 1983.

Franzwa, Helen H. "Pronatalism in Women's Magazine Fiction." In *Pronatalism: The Myth of Mom and Apple Pie*, edited by Ellen Peck and Judith Senderowitz. New York: Crowell, 1974.

———. "Female Roles in Women's Magazine Fiction, 1940–1970." In *Woman: Dependent or Independent Variable?* edited by Rhoda Kasler Unger and Florence L. Denmark. New York: Psychological Dimensions, 1975.

Friedan, Betty. *The Feminine Mystique*. New York: Norton, 1963.

Fussell, Paul. *Wartime: Understanding and Behavior in the Second World War*. New York: Oxford University Press, 1989.

Geise, L. Ann. "The Female Role in Middle Class Women's Magazines from 1955 to 1976: A Content Analysis of Nonfiction Selection," *Sex Roles* 5, no. 1 (1979): 51–62.

Goodwin, Doris Kearns. *No Ordinary Time: Franklin and Eleanor Roosevelt: The Home Front in World War II*. New York: Simon and Schuster, 1994.

Gould, Bruce, and Beatrice Blackmar Gould. *American Story: Memories and Reflections of Bruce Gould and Beatrice Blackmar Gould*. New York: Harper and Row, 1968.

Graebner, William. "Coming of Age in Buffalo: The Ideology of Maturity in Postwar America." *Radical History Review* 34 (1986): 53–74.

———. *The Age of Doubt: American Thought and Culture in the 1940s*. Boston: Twayne, 1991.

Gubar, Susan. " 'This is My Rifle, This Is My Gun': World War II and the Blitz on Women." In *Behind the Lines: Gender and the Two World Wars*, edited by Margaret Randolph Higgonet et al. New Haven: Yale University Press, 1987.

Halberstam, David. *The Fifties*. New York: Villard, 1993.

Hartmann, Susan M. "Prescriptions for Penelope: Literature on Women's Obligations to Returning World War II Veterans." *Women's Studies* 5 (1978): 223–39.

Harvey, Brett. *The Fifties: A Women's Oral History*. New York: HarperCollins, 1993.

Hatch, Mary G., and David L. Hatch. "Problems of Married Working Women as Presented by Three Popular Working Women's Magazines," *Social Forces* 37 (December 1958): 148–53.

Honey, Maureen. "Recruiting Women for War Work: OWI and the Magazine Industry during World War II." *Journal of American Culture* 3 (Spring 1980): 47–52.

Hooker, Richard J. *Food and Drink in America: A History.* Indianapolis: Bobbs-Merrill, 1981.

Horowitz, Daniel. "Rethinking Betty Friedan and *The Feminine Mystique:* Labor Union Radicalism and Feminism in Cold War America." *American Quarterly* 48, no. 1 (March 1996): 1–42.

Humphreys, Nancy K. *American Women's Magazines: An Annotated Historical Guide.* New York: Garland, 1989.

Jackson, Kenneth T. *Crabgrass Frontier: The Suburbanization of the United States.* New York: Oxford University Press, 1985.

Leman, Joy. " 'The Advice of a Real Friend': Codes of Intimacy and Oppression in Women's Magazines, 1937–1955." *Women's Studies International Quarterly* 3 (1980): 63–78.

Levenstein, Harvey. *Paradox of Plenty: A Social History of Eating in Modern America.* New York: Oxford University Press, 1993.

McCracken, Ellen. *Decoding Women's Magazines: From* Mademoiselle *to* Ms. New York: St. Martin's, 1993.

McDowell, Margaret B. "The Children's Feature: A Guide to the Editors' Perceptions of Women's Magazines." *Midwest Quarterly* 19, no. 1 (October 1977): 36–50.

Matthews, Glenna. *"Just a Housewife": The Rise and Fall of Domesticity in America.* New York: Oxford University Press, 1987.

May, Elaine Tyler. *Homeward Bound: American Families in the Cold War Era.* New York: Basic Books, 1988.

May, Lary, ed. *Recasting America: Culture and Politics in the Age of Cold War.* Chicago: University of Chicago Press, 1989.

Mayes, Herbert R. *The Magazine Maze: A Prejudiced Perspective.* Garden City, N.Y.: Doubleday, 1980.

Meyerowitz, Joanne. "Beyond the Feminine Mystique: A Reassessment of Postwar Mass Culture, 1946–1958." *Journal of American History* 79, no. 4 (March 1993): 1455–82.

———. *Not June Cleaver: Women and Gender in Postwar America, 1945–1960.* Philadelphia: Temple University Press, 1994.

Michel, Sonya. "American Women and the Discourse of the Democratic Family in World War II." In *Behind the Lines: Gender and the Two World Wars,* edited by Margaret Randolph Higgonet et al. New Haven: Yale University Press, 1987.

Milkman, Ruth. "American Women and Industrial Unionism during World War Two." *Behind the Lines: Gender and the Two World Wars,* edited by Margaret Randolph Higgonet et al. New Haven: Yale University Press, 1987.

Mullen, Bill, and Sherry Linkon, eds. *Radical Revisions: Rereading 1930s Culture.* Urbana: University of Illinois Press, 1996.

Nadel, Alan. *Containment Culture: American Narratives, Postmodernism, and the Atomic Age.* Durham: Duke University Press, 1995.

Okker, Patricia. *Our Sister Editors: Sarah J. Hale and the Tradition of Nineteenth-Century American Women Editors.* Athens: University of Georgia Press, 1995.

Robinson, Gertrude Jock. "The Media and Social Change: Thirty Years of Magazine Coverage of Women and Work (1950–1977)." *Atlantis* 3, no. 2 (Spring 1983): 87–111.

Shapiro, Laura. *Perfection Salad: Women and Cooking at the Turn of the Century.* New York: Farrar, Straus and Giroux, 1986.

Steinem, Gloria. "Sex, Lies, and Advertising." *Moving beyond Words,* 130–68. New York: Simon and Schuster, 1994.

Strasser, Susan. *Never Done: A History of American Housework.* New York: Pantheon, 1982.

Tebbel, John, and Mary Ellen Zuckerman. *The Magazine in America, 1741–1990.* New York: Oxford University Press, 1991.

Tuttle, William M. Jr. *"Daddy's Gone to War": The Second World War in the Lives of America's Children.* New York: Oxford University Press, 1993.

Walker, Nancy A. "Humor and Gender Roles: The 'Funny' Feminism of the Post–World War II Suburbs." *American Quarterly* 37, no. 1 (Spring 1985): 98–113.

White, Cynthia. *Women's Magazines, 1693–1968.* London: Michael Joseph, 1970.

Whittemore, Katharine, ed. *The World War Two Era: Perspectives on All Fronts from* Harper's *Magazine.* New York: Franklin Square Press, 1994.

Winship, Janice. *Inside Women's Magazines.* London: Pandora, 1987.

Wolseley, Roland E. *The Magazine World: An Introduction to Magazine Journalism.* New York: Prentice-Hall, 1951.

"Woman Power." *Newsweek,* March 30, 1970, 61.

Zuckerman, Mary Ellen. *Sources on the History of Women's Magazines, 1792–1960: An Annotated Bibliography.* New York: Greenwood, 1991.

Index

"Accessory after the Body"
   (*Mademoiselle*), 223–24
Adams, Clifford R., 121–25, 259
Adams, Henry, 164
Advertising, 61 (illus.), 98 (illus.), 144
   (illus.), 155 (illus.), 179 (illus.), 194
   (illus.)
   African American women and, 7
   amount of, 2
   articles as form of, 9
   consumer culture and, 5
   critiques of women's magazines and,
      232, 234–36
   dependence on revenue from, 14
   "expert" solutions offered by, 10–11
   fashion and, 208–09 (illus.), 215
   link between editorial content and,
      13–14
   technological developments and, 3
   women's roles and, 8
Advice columns, 2, 15
Advice-giving, 2
   child-rearing and, 99–100
   critique of, 259–61
   "expert" solutions offered by
      advertisements and, 10–11
   focus of magazines on, 9–10
African American women
   advertising and, 7
   magazines for, 7–8, 231
   working and, 64
Age of Doubt, The (Graebner), 16–17
Airline hostesses, 86–87
America, 254–57
America in Midpassage (Beard and
   Beard), 5–6
American Home, 155 (illus.)
American Magazine, 170–78
American Mercury, 234–42
American Story (Gould), 6
Anderson, Sherwood, 4

"Are American Moms a Menace?"
   (Scheinfeld; *Ladies' Home Journal*),
   100, 108–14
"Are Good Mothers 'Unfaithful' Wives?"
   (Evans; *Better Homes and Gardens*),
   102–6
"Are You Too Educated to Be a Mother?"
   (*Ladies' Home Journal*), 114–16
Atlantic Monthly, 229, 233, 242–46
"At My Age" (*Harper's Bazaar*), 220
Auden, W. H., 3, 19

Baber, Ray E., 112
Babies. *See* Children
Baumgartner, Leona, 90
Beard, Charles A., 5–6, 230
Beard, Mary Ritter, 5–6, 230
Beauty. *See* Fashion and beauty
Beecher, Catharine, 10
Belasco, David, 110
Berckman, Mary Godfrey and Fred, 38–44
Berry, Sally, 202–3
Better Homes and Gardens, 102–6
Bettleheim, Bruno, 100
Birth control, 99, 114–15
Birth rates, 12, 99, 114–16
Bok, Edward, 233
Bromley, Dorothy Dunbar, 34–35
Brown, Helen Gurley, 4
Bryan, Clark W., 14
Buck, Pearl S., 4, 19, 26–34
Byrne, Katharine M., 254–57

Cadden, Vivian, 166–70
Campbell, Levin H., Jr., 37–38
"Can This Marriage Be Saved?" series
   (*Ladies' Home Journal*), 97, 259–61
"Can You Date These Fashions?"
   (*Harper's Bazaar*), 199–201
Capote, Truman, 3
Carnegie, Mrs. Dale, 126–36

Carnegie, Hattie, 204
Carter, John Mack, 11, 228
Cather, Willa, 7
Catholic Church, 41–42, 97, 112, 254
Census Bureau, 63, 114
Charm, 250
Child, Lydia Maria, 10
Childbearing, 15, 153
Children
  advice on raising, 99–100, 177, 232
  juvenile delinquency and, 16, 18, 24, 45,
    49–50, 99
  marriage and motherhood issues and,
    102–6
  raising sons, 99–100, 108–114
  television watching by, 187–88
  working mothers during World War II
    and, 24, 39, 41, 42, 43, 44–56
  working women and care of, 50–56, 64,
    66, 73–74, 82, 84, 85
Children's Bureau, 53
Circulation, 1–2, 4, 233
Civilian Defense Volunteer Offices, 16
Civil rights movement, 7–8, 11
Civil War, 13
Class differences, 64
Cleaning. *See* Homemaking
Clothing. *See* Fashion and beauty
Cohn, David L., 6
Cold war, 8, 17
Colette, 3
College students, 3, 9
Colton, Jennifer, 82–85
Conover, Harry, 207
Consumerism
  advertisements and, 5
  economic changes and rise of, 12–13
  fashion and beauty and, 193–95
  postwar emphasis on, 17
Cooking. *See* Homemaking
Coontz, Stephanie, 18
Coronet, 64
  articles in, 126–36, 136–41, 195
Cosmopolitan, 4
Critiques, 228–61
  background to, 228–29
  "Happy Little Wives and Mothers"
    (Byrne; *America*), 254–57
  "The Magazines Women Read"
    (Griffith; *American Mercury*), 234–42
  "Marriage à la Mode" (Didion;
    National Review), 258–61
  "Up the Ladder from *Charm* to *Vogue*"
    (McCarthy; *Reporter*), 247–54

"What Every Woman Knows by Now"
  (Laski; *Atlantic Monthly*), 242–46
"The Women's Magazines"
  (Schlesinger; *New Republic*), 229–33
Culture
  during the 1950s, 17
  range of articles published on, 2, 6
Current Opinion, 228
Curtis, Cyrus H. K., 233

Dahl-Wolfe, Louise, 207–10
Day care centers, 55–56
Delafield, Ann, 223–24
Delineator, 13, 229
"Diary of Domesticity" feature (Taber;
  Ladies' Home Journal), 146
Didion, Joan, 228, 258–61
Diets, 195, 211–15, 223–25
"Diet Your Way to Beauty and Health"
  (*Coronet*), 195
Disney, Dorothy Cameron, 259
Divorce, 78, 97–99
"Do You Make These Beauty Blunders?"
  (Berry; *Good Housekeeping*), 202–3

Ebony, 7
Economic conditions, and consumer
  culture, 12–13
Edison, Thomas A., 145
Editors
  critics of, 6
  "guest editors" of *Mademoiselle*, 9
  range of articles published and, 2
  relationship between readers and, 4–5,
    6–7
Educational system
  articles on, 2, 8, 230
  motherhood and, 114–16
Eisenhower, Mrs. Dwight D., 132
"Electric Mixers — Strong Right Arms"
  (Kendall; *Good Housekeeping*),
    165–66
Employment. *See* Workplace
Engel, Janet, 211–15
Equal Rights Amendment, 64, 76–81
Essence, 7
Evans, Wainwright, 102–6
Exercise, 196–98, 227

Families
  birthrate changes and size of, 12
  care of, during World War II, 50–56
  Franklin Roosevelt on importance of, 18
  idealized 1950s image of, 17–18

juvenile delinquency and, 16, 18, 24, 45,
    49–50, 51, 99
motherhood role and, 99
returning soldiers and, 56–62
working mothers during World War II
    and, 37–56
working women and, 64
Fan magazines, 4
Farmer, Fannie, 169
Fashion and beauty, 193–95, 194 (illus.),
    208–9
advice on, 2
articles as a form of advertising for, 9
consumerism and, 193–95
critiques of women's magazines and,
    234–35, 242–46, 247–52
diets and, 195, 211–15, 223–25
magazines focused on, 3, 193
marriage and, 100–102
texts of articles on, 193–227
women soldiers and, 35–37
World War II and, 17
younger readers and, 3
"Fathers Are Parents Too" (*Woman's
    Home Companion*), 100
"Fattest Girl in the Class, The" (Engel;
    Seventeen), 211–15
Federal Bureau of Investigation, 17, 44, 45
"Feeding Five on $14 a Week" (*Good
    Housekeeping*), 13
Feminine Mystique, The (Friedan), 8, 11,
    12, 146, 228
Feminism, 11, 76, 78, 228
Fenichel, Otto, 111
Ferguson, Marjorie, 5
Fiction
critique of, 238–42
in women's magazines, 3, 4, 7, 8–9,
    18–19, 232
Fisher, Dorothy Canfield, 149–51
Fisher, M. F. K., 3, 156–61
Food and Drug Administration, 153
Forever Feminine (Ferguson), 5
Foster, Steven, 110
Foster care, 53–55
"Fraud of Femininity, The" (Friedan;
    McCall's), 12
Frederics, John, 206
Friedan, Betty, 8, 11, 12, 146, 228

Gallup poll, 12
Gender issues, and women's magazines,
    5, 6
Generation of Vipers, A (Wylie), 99

Gilbreth, Lillian, 95
Glamour, 3–4, 250
Glamour of Hollywood, 4
Glasgow, Ellen, 7
Godey's Lady's Book, 6, 9, 13, 193, 229
Good Housekeeping, 98 (illus.), 258
advice-giving in, 10
articles in, 13, 17, 19, 35–37, 65–71,
    82–85, 97–99, 100–102, 145, 152–54,
    165–66, 196–98, 202–3
circulation of, 2
cover price of, 2
critique of, 259
homemaking articles in, 145
link between advertising and editorial
    content and, 13–14
readers of, 3, 7
School for Brides in, 97, 100
"Women in Politics" series in, 8
working women and, 64
during World War II, 16, 17
Good Housekeeping Institute, 14, 145
Gould, Beatrice Blackmar, 6, 12, 14
Gould, Bruce, 6, 7, 12, 14
Graebner, William, 16
Grafton, Samuel, 230
Graham's, 229
"Granny's on the Pan" (Whitbread and
    Cadden; *Redbook*), 166–70
Griffith, Ann, 234–42
"Guest editors" of *Mademoiselle*, 9

Hair care. *See* Fashion and beauty
Hale, Sarah Josepha, 6, 9, 10, 13
Halsey, Margaret, 145
Hamilton, Alice, 76–81
"Happy Little Wives and Mothers"
    (Byrne; *America*), 254–57
Hardy, E. J., 132
Harper's, 233
Harper's Bazaar, 250
advertising in, 2
articles as a form of advertising in, 9
articles in, 156–61, 198–99, 199–201,
    224–25
fashion focus of, 3, 193, 248, 253–54
working women and, 64
Hearst, William Randolph, 14
Henderson, Leon, 39
Hollywood fan magazines, 4
Homemaking, 144 (illus.), 145–46, 179
    (illus.)
critiques of women's magazines and,
    235, 236–38

Homemaking *(cont.)*
    Eleanor Roosevelt on, 70
    magazines' promotion of, 8
    text of articles on, 145–92
Homeward Bound (May), 18
Homosexuality, 111
Honey, Maureen, 16
Hoover, J. Edgar, 17, 44–47
Household management. *See*
    Homemaking
"Housekeeping Need Not Be Dull"
    (Fisher; *Ladies' Home Journal*),
    149–51
"How America Lives" series (*Ladies'
    Home Journal*), 7, 64
Howard, Sidney, 109
"How a Woman Should Wear a Uniform"
    (*Good Housekeeping*), 35–37
"How to Help Your Husband Get Ahead"
    (Carnegie; *Coronet*), 126–36
"How to Look Halfway Decent" (Smith;
    McCall's), 15, 225–27
"How to Stay Married Though Unhappy"
    (Sheen; *Good Housekeeping*), 97–99
Husbands. *See* Men

"If You Ask Me" column by Eleanor
    Roosevelt, 19
"I Gave Up My Career to Save My
    Marriage" (*Ladies' Home Journal*), 64
Infants. *See* Children
"Is There a Plot against Women?" (Jones;
    Ladies' Home Journal), 180–82

Jersild, Arthur T., 113
Jones, Paul, 180–82
Juvenile delinquency, 16, 18, 24, 45,
    49–50, 51, 99

Kendall, Helen W., 165–66
Khrushchev, Nikita, 18
Kiviette, 206
Knowlton, Robert J., 170–78

Labor. *See* Workplace
Ladies' Home Journal, 179 (illus.), 258
    advertising revenue and, 14
    advice-giving in, 10
    articles in, 19, 24–25, 26–34, 37–44,
        56–62, 63, 64, 71–75, 76–81, 99, 100,
        108–14, 114–16, 121–25, 145, 146–48,
        149–51, 161–64, 180–82
    "Can This Marriage Be Saved?" series
        in, 97

child-rearing advice in, 99, 100
circulation of, 1–2, 233
critiques of, 228, 229, 230, 231, 233
editors of, 6, 12
homemaking articles in, 145–46
"How America Lives" series in, 7, 64
"Making Marriage Work" series in, 13,
    97
marriage and motherhood articles in,
    97, 99
readers of, 3
response of readers to, 11
"Sub-Deb" column for teenagers in, 2
Taber's "Diary of Domesticity" feature
    in, 146
"What Do the Women of America
    Think?" series in, 14–15
working women and, 63
during World War II, 16
Landis, Judson T., 140
Landis, Paul H., 136–41
Lane, Gertrude, 6–7
Laski, Marghanita, 242–46
"Lass with the Delicate Air, The"
    (*Mademoiselle*), 215–20
Laws and legislation, and working
    women, 78–81
League of Nations, 28–29
Lenroot, Katharine F., 51
Leopold, Alice K., 94
"Let's Be Realistic about Divorce"
    (*Ladies' Home Journal*), 99
Levy, David M., 109–10
Lewis, Sinclair, 7
Life, 7, 18
Lincoln, Abraham, 110
"Line Forms Here, The" (*Mademoiselle*),
    221–22
"Lively Art of Eating, The" (Fisher;
    Harper's Bazaar), 156–61
Look, 7
Lord, Mary Pillsbury, 92
Love in America (Cohn), 6
Lubold, Joyce, 189–92
Lynd, Helen, 232–33
Lynd, Robert, 232–33
Lynes, Russell, 3

McCall's, 250, 258
    articles in, 12, 15, 141–43, 182–88,
        189–92, 225–27
    Betsy McCall paper doll in, 2
    circulation of, 2, 233
    consumer culture and, 13

critiques of, 229, 230, 231–32, 259
editors of, 6
Eleanor Roosevelt's series in, 19
fashion focus of, 193
"togetherness" theme of, 18
"Washington Newsletter" feature of, 231
working women and, 6
McCarthy, Mary, 228, 247–54
McCullers, Carson, 3
McGinley, Phyllis, 100
Mademoiselle
advertising in, 9
articles in, 8, 63–64, 86–87, 215–20, 221–22, 223–24
Auden's article in, 19
fashion focus of, 3, 193, 251–52
"guest editors" of, 9
Magazine Bureau, 15–16
Magazines. *See* Women's magazines; *individual magazines*
"Magazines Women Read, The" (Griffith; *American Mercury*), 234–42
Magazine War Guide, 16, 23
Make-up. *See* Fashion and beauty
"Making Less of Yourself" (*Harper's Bazaar*), 195, 224–25
"Making Marriage Work" (Adams; *Ladies' Home Journal*), 13, 97, 121–25
Male and Female (Mead), 19, 139–40
Mangone, Philip, 206–7
Mann, Thomas, 19
Manpower Inc., 90, 91
Marble, Alice, 34
"Marriage à la Mode" (Didion; *National Review*), 258–61
Marriage and motherhood, 97–100
Buck's article on, 19
critique of focus of magazines on, 254–57
dissatisfaction in, 116–21
focus on children and, 102–6
number of articles on, 8
raising sons, 99–100, 108–14
readers and, 3
sexual relations in, 119–120, 123, 136–41
soldiers and, 56–62, 106–7
text of articles on, 97–143
working women and conflicts regarding, 64, 71–75
"Married Woman Goes Back to Work, The" (*Woman's Home Companion*), 64, 87–96
Maugham, Somerset, 4
May, Elaine Tyler, 18

Mayes, Herbert R., 6, 7
Mead, Margaret, 19, 139–40
"Meet the Berckmans: The Story of a Mother Working on Two Fronts" (*Ladies' Home Journal*), 37–44
Men. *See also* Gender issues, and women's magazines; Soldiers
advice articles by, 10
as editors of women's magazines, 228
fatherhood and, 100, 104–5, 112, 113
in fiction, 240
focus of wives on career of, 126–36
homemaking and, 170–78, 180–82
sexual relations in marriage and, 139–40
working mothers during World War II and, 38–44
working women and family issues and, 75, 94–95
Michel, Sonya, 18
Middle class, 13, 17
Miles, Catharine C., 111
Military. *See* Soldiers
Minority groups, 231. *See also* African American women
Monroe, Marilyn, 17
"Mood Has Changed, The" (*Harper's Bazaar*), 198–99
Moss, Louise, 54
Motherhood. *See* Marriage and motherhood
"Mothers . . . Our Only Hope" (Hoover; *Woman's Home Companion*), 44–47
Murray, Arthur and Kathryn, 132–33
Murrin, Ruth, 100–102
"My Husband Says — " (Welshimer; *Good Housekeeping*), 100–102
"My Love Affair with the Washing-Machine Man" (Lubold; *McCall's*), 189–92

Nation, 233
National Business Woman, 3
National Consumers League, 77
National Review, 258–61
New Republic, 229, 229–33
Nixon, Richard, 18
Novels, 3, 4, 18–19

"Occupation — Housewife" (Thompson; *Ladies' Home Journal*), 161–64
Office of Civilian Defense, 54
Office of War Information Magazine Bureau, 15–16
"103 Women Sound Off" (Robinson; *McCall's*), 182–88

Opinion polls
 women's magazines and, 4–5, 14–15
 on working women, 63
 during World War II, 24–25

Parents, 100
Peck, Bernice, 223–24
Pennock, Grace L., 146–48
Perkins, Frances, 69–70
Personal appearance. *See* Fashion and
 beauty
Peterson's, 229
Pfizer, Beryl, 141–43
Phelps Publishing Company, 14
Pictorial Review, 13, 229, 250
"Pious Pornographers, The" (Williams;
 Playboy), 228
Playboy, 17, 228
Poetry, 2
Politics, 8, 13, 65–71, 230, 231
Polls. *See* Opinion polls
Pope, Elizabeth, 204–10
Popenoe, Paul, 259–61
Porter, Katherine Anne, 3
Powers, John Robert, 205
Priest, Ivy Baker, 91
Propaganda, 16
Public opinion. *See* Opinion polls

"Question-Box, The" (Sherwin; *Good
 Housekeeping*), 152–54

"Race Suicide of the Intelligent"
 (Thompson; *Ladies' Home Journal*),
 99
Racism
 advertising and, 7
 magazine articles and, 8
Read, Ruth Anna, 196–98
Reader's Digest, 256
Redbook
 articles in, 11, 106–7, 166–70, 204–10
 circulation of, 2
 response of readers to, 11
 "What's on *Your* Mind?" column in, 11
 during World War II, 23
Reporter, 229, 247–54
Revolution, 3
Rinehart, Mary Roberts, 97
Robinson, Selma, 182–88
Roles of women
 advertising and, 8
 articles on, 2, 16–17
 working mothers and, 24, 83–84

Roman Catholic Church, 41–42, 97, 112,
 254
Roosevelt, Eleanor (Mrs. Franklin D.), 8,
 65–71, 231
Roosevelt, Franklin Delano, 16–17, 18, 110
Rubinstein, Helena, 224–25

Saturday Evening Post, 100
Schauffler, Goodrich C., 259
Scheinfeld, Amram, 100, 108–14
Schlesinger, Elizabeth Bancroft, 229–33
School for Brides, *Good Housekeeping*, 97,
 100
"Scrambled Housewife, The" (*Ladies'
 Home Journal*), 145
Service magazines, 2–3, 63, 258
Seventeen
 appeal to young readers of, 3–4
 articles in, 23, 63, 211–15
 critique of, 252–53
 fashion focus of, 193
 "Sex, Lies, and Advertising" (Steinem),
 14
Sheen, Fulton J., 97–99
Sherwin, Carl P., 152–54
"Shopping in Wartime for the Men of the
 House" (*Good Housekeeping*), 17
Short stories, 2
"Should We Draft Mothers?" (Wood;
 Woman's Home Companion), 48–50
"Six Rude Answers to One Rude
 Question" (Pfizer; *McCall's*), 141–43
Smith, Elinor Goulding, 225–27
Social class differences, 64
Social conditions, 11–19
 cold war and, 8
 consumer culture and, 17
 critiques of women's magazines and
 reporting on, 230–33, 238
 idealized family of the 1950s and, 17–18
 issues of race and, 7–8
 juvenile delinquency and, 16, 18, 24, 45,
 49–50, 51, 99
 opinion polls on, 14–15
 role of women's magazines in
 reporting, 6
 wider world reported in magazines
 and, 18–19
 women's movement and, 11–12
 women's rights and, 2–3
 working mothers during World War II
 and, 24, 39, 41, 42, 43, 44–50
 working women and, 3, 11, 12
 World War II efforts and, 15–17

Soldiers
  children of wives of, 52–53
  marriages and, 106–7, 111
  return of, and families, 56–62
  women as, 35–37
South, readers in, 7, 64
Soviet Union, 8, 18
Spock, Benjamin, 10, 99
Stafford, Jean, 3
Stardom, 4
"Starting from Scratch" (Pennock; *Ladies' Home Journal*), 146–48
Steinbeck, John, 19
Steinem, Gloria, 14
Strecker, Edward A., 100, 108–9
"Sub-Deb" column, *Ladies' Home Journal*, 2
Subscription prices, 2
Suburbs, 11, 13, 18
Suffrage, 3
Surplus Marketing Administration, 35

Taber, Gladys, 146
Tan, 7
Technology
  advertising and, 3
  social transformation and, 11
Teenagers
  advice column for, 2
  magazines aimed at, 3–4, 252–53
Television, 2, 13, 17, 187–88
Temporary employment, 90–91
Terman, Lewis M., 111, 128, 139
This Demi-Paradise (Halsey), 145
Thompson, Dorothy, 13, 99, 146, 161–64, 231
"Those Simple Little Exercises" (Read; *Good Housekeeping*), 196–98
Time, 233
Toombs, Alfred, 50–56
Training, 35
Truman, Harry, 110
"Two Ways to Dress in Wartime" (*Woman's Home Companion*), 17

Una, 3
"Unchastity Is a Sin" (McGinley; *Good Housekeeping*), 100
"Up the Ladder from *Charm* to *Vogue*" (McCarthy; *Reporter*), 247–54

Van Buren, Abigail, 2
Vogue, 3, 247–50, 253–54
Voice of Industry, 3
Volunteer work, 16, 23, 35, 64

Walsh-Healy Wage and Hour Law, 79
War Advertising Council, 15, 232
"War Babies" (Toombs; *Woman's Home Companion*), 50–56
War Bonds and Stamps, 16, 39, 41, 44, 232
War Manpower Commission, 53
Warner, W. Lloyd, 131
"Washington Newsletter" feature (*McCall's*), 231
Waugh, Alec, 4
Way We Never Were, The (Coontz), 18
Welfare organizations, 53–55
Wells, Carveth, 127
Wells, Zetta, 127
Welshimer, Helen, 100–102
West, Jessamyn, 3
"What Are You Doing about the War" (*Seventeen*), 23
"What Do the Women of America Think" series (*Ladies' Home Journal*), 14–15
"What Do the Women of America Think about War?" (*Ladies' Home Journal*), 24–25
"What Every Woman Knows by Now" (Laski; *Atlantic Monthly*), 242–46
"What Is a 'Well-Dressed' Woman?" (Pope; *Redbook*), 204–10
"What Is 'Normal' Married Love?" (Landis; *Coronet*), 136–41
"What Makes Wives Dissatisfied?" (*Woman's Home Companion*), 116–121
"What's on *Your* Mind?" column (*Redbook*), 11, 106–7
What's Wrong with American Mothers? (Strecker), 100
"When Your Soldier Comes Home" (*Ladies' Home Journal*), 56–62
Whitbread, Jane, 166–70
White House Conference on Children in a Democracy, 18
"Why I Am *against* the Equal Rights Amendment" (*Ladies' Home Journal*), 76–81
"Why I Quit Working" (Colton; *Good Housekeeping*), 82–85
Wiley, Harvey W., 14
Williams, Ivor, 228
Wilson, Woodrow, 28–29
Winter, Elmer, 91
Wives. See Marriage and motherhood
"Woman of the Future, The" (Edison; *Good Housekeeping*), 145

"Womanpower in the Election" (Beard and Beard; *Woman's Day*), 230
Woman's Advocate, 3
Woman's Day, 230, 231, 232, 233
Woman's Home Companion, 144 (illus.), 194 (illus.)
    advice-giving and advertisements in, 11, 12
    articles in, 17, 34–35, 44–47, 48–50, 50–56, 64, 87–96, 100, 111–21
    circulation of, 2, 233
    cover price of, 2
    critiques of, 229, 230–31, 232
    editorial staff of, 5, 6–7
    marriage articles in, 97
    "reader-editors" of, 15
    working women and, 12, 63
Woman's Journal, 3
Woman's Party, 76, 78
"Women and War" (Buck; *Ladies' Home Journal*), 26–34
"Women in Flight" (*Mademoiselle*), 86–87
"Women in Politics" series (Roosevelt; *Good Housekeeping*), 8, 65–71
Women's magazines, 1–19
    advertisements in. *See* Advertising
    advising role of, 9–11
    for African American women, 7
    celebrating women's achievements in, 8–9
    circulation of, 1–2, 4
    consumer culture and, 12–13, 17
    critics of, 5–6
    defining women's aspirations in, 4–5, 15
    editors of. *See* Editors
    fiction published in, 3, 4, 7, 8–9, 18–19
    idealized family of the 1950s and, 17–18
    issues of race and, 7–8
    mix of elements in, 2
    opinion polls reported in, 14–15
    response of readers to, 11
    role of, 1–11
    service magazines and, 2–3
    society and women reflected in, 11–19
    wider world reported in, 18–19
    women's movement and, 11–12
    women's rights and, 2–3
    working women and, 3, 11, 12
    World War II efforts and, 15–17
    for young adults, 3–4
"Women's Magazines, The" (Schlesinger; *New Republic*), 229–33
Women's movement, 11–12

Women's rights, 2–3
Women's roles. *See* Gender issues; Roles of women
Women workers. *See* Workplace
"Women Work for Their Country" (Bromley; *Woman's Home Companion*), 34–35
Wood, James Madison, 24, 44–45, 48–50
Woodhull and Claflin's Weekly, 3
Woolf, Virginia, 164
"Working Wives Make the Best Wives" (*Ladies' Home Journal*), 63
Working women. *See* Workplace
Workplace, 63–64
    advice for, 9, 11, 12
    articles on, 8, 16
    attitudes toward, 12
    children of, 50–56, 64, 66, 73–74, 82, 84, 85
    family issues and, 64
    focus of magazines on, 63–64
    juvenile delinquency and, 16, 18, 45, 49–50, 51
    labor legislation affecting, 78–81
    magazines for, 3, 63–64
    marriage and, 64, 71–75
    reasons for quitting work, 82–85
    reasons for returning to work, in the 1950s, 88–90
    text of articles in, 71–96
    World War II and, 23, 37–44, 50–56
World War II
    attitudes toward working women during, 12
    background to, 23–24
    children and absentee mothers during, 44–50
    fashion during, 3, 35–37
    support for war effort during, 15–16, 23–24
    text of articles concerning, 24–62
    women workers during, 23, 34–35, 37–56
Writers' War Board, 15
Wylie, Philip, 4, 99

"You Can't Have a Career and Be a Good Wife" (*Ladies' Home Journal*), 63, 71–75
Young adults
    advice column for, 2
    magazines aimed at, 3–4
"Your Wife Has an Easy Racket!" (Knowlton; *American Magazine*), 170–78

Printed in the United States
By Bookmasters